Mathematics of Environmental Processes

Mathematics of Environmental Processes

Thomas D. Waite
Northwestern University

Neil J. Freeman
University of Miami

Lexington Books
D. C. Heath and Company
Lexington, Massachusetts
Toronto

Library of Congress Cataloging in Publication Data
Waite, Thomas D
Mathematics of environmental processes.
Includes index.
1. Environmental engineering—Mathematics. I. Freeman, Neil J., joint author. II. Title.
TD153.W28 510.2'4628 76-25770
ISBN 0-669-00979-2

Published simultaneously in Canada

Printed in the United States of America

International Standard Book Number: 0-669-00979-2

Library of Congress Catalog Card Number: 76-25770

To the Kids: Oliver, Whitney, Cindi, and Keith

Contents

List of Figures

List of Tables

Preface

The definition, and subsequent quantification of environmental processes has progressed to the point where many processes can now be accurately modeled. Most modern environmental texts now include mathematical representations of unit processes so that better design criteria may be developed. The result of this evolution in sophistication has been an increased demand on the student's mathematical ability. While most engineers and scientists have been introduced to basic math, recalling special treatments for certain solutions is a time-consuming endeavor, due mostly to the general nature of text books in the mathematics area.

This book attempts to condense the mathematical techniques necessary for treating most environmental models. This is possible because most processes involved in environmental situations can be modeled by a relatively few mathematical functions. It therefore is important for the student to understand these functions and be able to treat them easily. The best example of this type of function is the exponential. While this function appears in many models, its behavior is usually not fully appreciated, nor is its differentiation or integration. We have presented an entire chapter which, while being a review of calculus, thoroughly describes the different types of functions, ending with the exponential.

Another useful aspect of this book is the treatment of the mathematics involved in material transport. The most fundamental aspect of any environmental process is the phenomenon of mass transport and hence should be completely understood. Once again, due to the generality of many material transport texts, it is difficult to follow step by step through the mathematical derivations. Chapter 3 presents a clear but rigorous treatment of material transport. The reader should be able to follow logically the derivation of a fundamental relationship that includes advective and diffusive transport. Relationships of this type are the foundation of chapter 4, which deals with the fabrication of environmental models.

The focal point of the book is chapter 5, "Treating Differential Equations." Here we first describe how to "key out" the type of differential equation that is to be treated. Because most common models are simple differential equations, solution procedures are relatively straightforward. However, if the equation is not mathematically simple then solution procedures may be very complex if any solution exists at all. Therefore, one of the most important facets of handling differential equations is being able to recognize if the equation being treated can be solved directly, or if advanced numerical techniques are required for an approximate solution.

Chapter 5 describes the procedures for solving certain differential equations. Examples of mathematically-complex equations are shown, utilizing environmental situations. This should give the reader an idea of how to select functions for certain processes when more than one could be used to minimize mathematical complexity. In addition, the environmental models that were created in chapter 4 are treated and used as examples in the text. Thus a continuity is maintained through the book, in that several environmental models are created from known chemical or biological phenomena, and then treated in chapter 5 to yield the integrated form of the rate expression.

The last chapter of this book deals with statistical analysis of environmental data. Common methods are outlined for characterizing data and determining a probability law for environmental variables. The last section of the chapter describes concepts to be used in experimental setups, including certain experimental designs for maximum reliability of test results.

This book should find wide use with a broad spectrum of environmental courses. It can be used as a supplementary text for any graduate environment modeling, waste treatment, or water quality course. It can also be used at the advanced undergraduate level for students who have had differential and integral calculus.

We are grateful to Dr. Earl Rich for contributing chapter 6 to our book. Gratitude is also extended to Dr. Murray Mantell for many helpful suggestions, and Ms. Sara Brittle and Mrs. Kathryn Waite for typing and reviewing the manuscript.

Mathematics of Environmental Processes

1 Introduction

Environmental engineering is a relatively new discipline of engineering. Combining the classical fields of sanitary engineering, physical chemistry, quantitative biology, and physics, this new area has developed into a valuable discipline for coping with environmental problems. The quantitative nature of this new field has also been developing very rapidly. Original design procedures for waste water treatment facilities, for instance, were based on scientific phenomena which had been described years ago. The young engineer in training was not usually required to understand the origin of all his design relationships, so he had very little insight into the limitations of his product. Recent concern over the lack of new design processes for waste treatment has brought about the necessity for better understanding of the fundamentals behind each design method. Modern texts such as that edited by Weber [1] emphasize the derivation of mathematical relationships (models) governing unit operations in water and waste treatment. This type of approach places a more rigorous demand on students for the understanding and use of their mathematical tools. It is the purpose of this text to present a concise review of the mathematical tools most useful in the study of environmental processes. We assume the reader has previously been introduced to undergraduate level mathematics or is a serious newcomer.

Systems Analysis Approach

Environmental systems besides those used for water and waste water treatment must also be mathematically modeled to accomplish rational resource planning. Predictive results of water quality changes, for instance, due to modifications in watershed areas, can only be made if accurate, functional models exist which represent the situation in question. In these cases modelers attempt to dissect complex phenomena into understanding subsets. These subsets can then be investigated mathematically to observe their relative contribution to the whole. Environmental models are not only written to evaluate quality changes that may occur due to man's activities, but also to provide the basis for economic analysis. For example, a model could be developed which represented the movement of phosphorus through a system (including biological and physical-chemical interactions). Utilizing this model, a community could evaluate its input to the phosphorus pool via sewage discharges and surface runoff. A decision could then be made concerning the necessity for initiating a costly tertiary treatment process for their sewage, or perhaps diversion of runoff waters. Thus it is possible for the engineer and administrator to evaluate the environmental significance of a project as well as the cost/benefit ratio.

It should be apparent to the reader that the systems analysis approach is the framework for the development of most environmental models. These mathematical models allow for the making of rational decisions concerning environmental modifications.

Water Quality Models

A good example of systems analysis development in the environmental area is seen in the evolution of water quality models. The origin of the mathematical representation of water quality was the dissolved oxygen model, initially conceived by Streeter and Phelps [2], depicting the resultant water quality of the Ohio River after receiving sewage discharges. The original model was a linear differential equation containing only a reaeration component and biochemical oxygen demand (BOD) component. Through the years, more terms were added to the original model to better predict dissolved oxygen concentrations in river systems. As the model grew, so did the mathematical complexity.

Recently the art of modeling environmental processes has become more sophisticated. Models have been generated for river systems, lakes, estuaries, oceans, and even the world. Many of these models deal with process rates and include functions that are difficult to analyze mathematically. A recent text by Thomann [3] displays some of the mathematical sophistications that can be found in environmental models. While the techniques required to solve some of the models are beyond the mathematical tools given to engineers and scientists at the undergraduate level, a large percentage of useful environmental models are represented by first-order, linear, constant coefficient, differential equations. This type of differential equation has usually been treated by undergraduate students in their mathematics courses, but students tend to have difficulty identifying and applying the solution methods to these equations when they model environmental systems.

Use of Rate Laws

In an effort to describe the many states of a system, we are often faced with a tremendously complex task. To avoid attempting to establish a relation between all states of a process, we can investigate a single relation between one state and the next. This type of law expresses how a certain state will develop in the near future. This law is usually referred to as the "law of causality" and allows us to describe mathematically natural processes in terms of infinitesimal increments of time and space—that is, in terms of rates of change in state.

Rates of processes are usually expressed as derivatives—that is, some change with time—and relations among these derivatives and other functions are differential equations. Thus differential equations express relationships between increments of certain quantities and the quantities themselves. From the above discussion it appears that a differential equation is a good representation of the law of causality.

With an understanding of the above concepts, it becomes apparent that we can describe certain states of a process by examining only the rate of their progression. Thus almost all environmental models are based on rates of the processes entering into the analysis. For example, the process of a gas diffusing into a liquid can be defined by the rate of gas transfer to the liquid, that is

$$\frac{dC}{dt} = k(C_s - C)$$

where dC/dt = change in gas concentration with time

k = rate constant

C_s = saturation concentration of gas

C = concentration of gas at any time (t)

Certain mathematical functions appear in many models of environmental processes. Perhaps the most commonly utilized function is the exponential function (a^x). Many biological and chemical phenomena appear to behave in an exponential manner; thus this function has appeared in most models. Understanding the mathematical behavior of the exponential function is basic to understanding most environmental phenomena. As is shown in the next chapter, large numbers are generated very rapidly when a process is proceeding in proportion to the amount of material already present—that is, exponentially increasing.

The following chapters attempt to define and analyze those functions and differential equations that are most often encountered in environmental models. After following through this text, the student should be able to treat most of the common differential equations and functions used in water quality management and general environmental models.

References

1. Weber, W. J. (1972), *Physiochemical Processes for Water Quality Control*, New York, Wiley-Interscience.
2. Streeter, H. W., and E. B. Phelps (1925), *U.S. Pub. Health Bull.* 146.
3. Thomann, R. V. (1972), *Systems Analysis and Water Quality Management*, New York, McGraw-Hill.

2 Elements of Calculus

Basic Concepts

Any reaction or process that increases or decreases by a constant percentage of the whole in a constant time period is behaving in an exponential manner. Linear processes are those which change by a constant amount in a constant time period. An example of linear growth would be the addition of one leaf per day on a plant stem. This addition is constant, and does not depend on the number of leaves already present. Exponential reactions, however, depend on the amount of reacting material present. Biologically, the best representation of exponential growth can be witnessed by the reproduction of unicellular microorganisms. One cell produces two, two produce four, four produce eight, etc. This population is growing in an exponential manner, such that:

$$\text{after 1 generation, population} = 2 = 2^1$$

$$\text{after 2 generations, population} = 2 \times 2 = 4 = 2^2$$

$$\text{after 3 generations, population} = 4 \times 2 = 8 = 2^3$$

$$\text{after 4 generations, population} = 8 \times 2 = 16 = 2^4$$

or the population size P after n generations is given by

$$P - 2^n \tag{2.1}$$

In this particular example, the population doubles after each generation. Other exponential processes may not be founded on a base two system, so doubling times must be computed in another manner.

Equation (2.1) illustrates that processes reacting in an exponential manner increase or decrease very rapidly. An example of what exponential growth of bacteria represents can be seen by calculating the production of bacterial cells assuming exponential growth continues for just two days. The model bacterium in the example will be the common *E. coli* which has a doubling time of about twenty minutes.

The population size will double every twenty minutes or 3 times per hour; thus the population size after forty-eight hours will be $2^{(3\times48)} = 2^{(144)} = 2.2(10)^{43}$ cells. Assuming the average volume of an *E. coli* cell to be 0.53

μ^3 with a density of 1.07 g/cm^3, the resultant weight of the bacterial population will be $2.2(10)^{43}(0.53)(1.07)(10)^{-12} = 1.25(10)^{31}$ grams $= 1.38(10)^{25}$ tons which is *2000* times the weight of the earth.

Of course environmental constraints prevent such unrestricted growth from continuing indefinitely, so these predicted values are purely academic.

The exponential function is often treated and occurs quite naturally in conjunction with the logarithmic function. The reader probably has encountered the logarithmic function first, in an algebra course, as a tool to simplify certain numerical calculations of arithmetic. These functions are of interest here in that they occur quite frequently in mathematics applied to environmental processes. The advance of a chemical reaction, bacterial growth, and radioactive decay are examples of such processes and are discussed in detail in later sections. The treatment of the logarithmic function and the exponential function in this chapter serves as a review of that given in elementary calculus. First, however, it is appropriate to review other functions so that the positions of the exponential and logarithmic functions in mathematical hierarchy can be determined.

Functions

A function is a law or rule which describes how one variable quantity is related to another. By variable quantity (or just variable) we mean a symbol, such as x or y, that can stand for various numerical values. Symbolically, the expressions

$$y = f(x),\ y = F(x),\ y = g(x)$$

or others similar to them state that y is a function of x—that is, for each value of x there corresponds a value of y according to some law. The symbol x is called the *independent* variable, and y the *dependent* variable, or x is called the *argument* of the function y. For example, if the variable g represents the increase in population size of an organism during time t according to a law G (the function) where t is the independent variable, then $g = G(t)$ (read g equals G of t). This is the rule of growth of the organism. Specifically consider the case where the law G states that at any specific time t, the increase in population size can be determined by raising the number two to the power t. In equation form this statement can be written as

$$g = 2^t \tag{2.2}$$

It is common usage to interchange the symbols G and $G(t)$ and refer to both as the function or law, although the latter is sometimes referred to, pedantically, as the value of the function G at time t.

Ordinarily, the independent variable is constrained within a certain interval of numbers, called the *range* of the independent variable; for example

$$a \leq x \leq b, \quad a < x < b, \quad a \leq x < b, \quad a < x \leq b$$

These expressions are equivalent to the statement that x can be any number within the interval of numbers from a to b. The absence of an equal sign, however, denotes that the adjacent number is excluded from the interval. Also, the dependent variable is usually constrained to a similar interval, called its *domain*.

A graphical representation of a function can be obtained by plotting x and corresponding values of y on a rectangular coordinate system. It is expected that the reader is familiar with the use of rectangular and perhaps other coordinate systems.

A function is said to be *single valued* if for each value of x there is only one assigned value of y, as for example

$$y = x$$

$$y = x^2$$

$$y = \sqrt{(1 - x^2)}$$

$$y = \cos x$$

If, however, there exists more than one value of y for each assigned value of x, the function is said to be *multiple valued*. Examples of equations where y is defined as a multiple-valued function of x are

$$x^2 + y^2 = 1 \text{ (circle of unit radius)}$$

$$y^2 = x$$

$$x = \cos y$$

Solving for y in the first example yields $y = \pm\sqrt{(1 - x^2)}$; thus it is seen that the equation defines y as a double-valued function of x. The equation $x = \cos y$ allows an infinite number of values of y for a particular value of x; for example, if $x = 1$, then $y = 0, \pm 2\pi, \pm 4\pi$, etc. satisfies the equation $1 = \cos y$. For clarity all functions will be considered to be single-valued functions in this text unless noted otherwise.

If a function always increases with an increase in the value of x, it is called a *monotonically increasing* function. If a function always decreases with an increase in the value of x, it is called *monotonically decreasing*. Examples of these types of functions are shown in figure 2–1, revealing the property of either a positive or negative slope at all points.

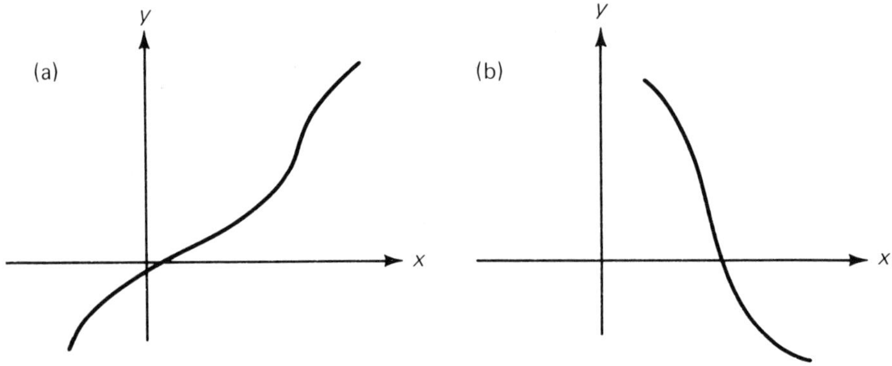

Figure 2–1. Monotonic Functions: (a) Monotonically Increasing; (b) Monotonically Decreasing

Graphs of the functions discussed thus far can be used to exhibit the property of continuity, a concept of basic importance in the discussion of functions. Basically, continuity of a function $y = f(x)$ implies that for small changes in x, there will be small changes in y. The graph of a continuous function is an unbroken curve. It is this basic property that is assumed when plotting data points, and a continuous curve is sketched through these points.

Since only continuous functions have been examined thus far, it is instructive to consider some examples of functions which are discontinuous. The most obvious cases of discontinuous functions are those that exhibit a step or jump in value. For example, the function defined by $f(x) = 0$ for $-\infty < x < 0$, and by $f(x) = 1$ for $0 < x < \infty$, is shown in figure 2–2. This function is commonly called the *unit-step function* and is used to mathematically describe an abrupt or instantaneous change to a system.

The function $y = 1/x$, illustrated in figure 2–3, exhibits an infinite discontinuity because the function tends to $\pm\infty$ in the vicinity of the origin—that is $y \to 1/0$. Any infinity of a function, positive or negative, defines a point of discontinuity.

Differentiation

Due to the dynamic nature of the systems and processes that are of interest throughout this text, a review of differentiation (forming a derivative), which examines the rate of change of a function, is indispensible. In calculus the derivative is treated through the concept of limits. This treatment usually discusses two important interpretations of the derivative; (a) the derivative of a function $y = f(x)$ at any point of a curve is equal to

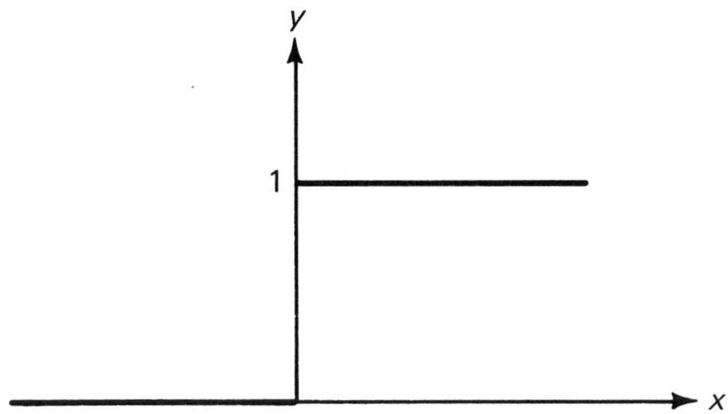

Figure 2–2. Discontinuous Step Function

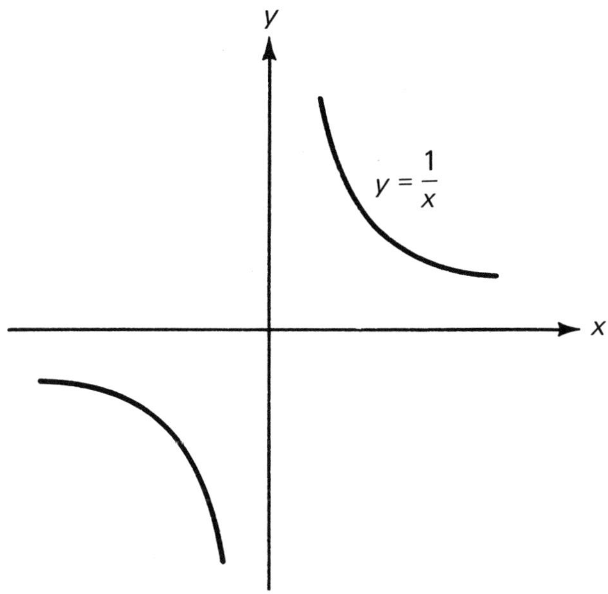

Figure 2–3. Plot of Function $y = 1/x$

the slope of the tangent line to the graph of the function at that point—that is, the derivative of a function is the slope of the graph of the function (see figure 2–4); (b) the derivative of a function is equal to its instantaneous rate of change. This last interpretation is of particular importance because, as is shown later, certain models describing processes such as bacterial growth are developed according to the observation that the rate of

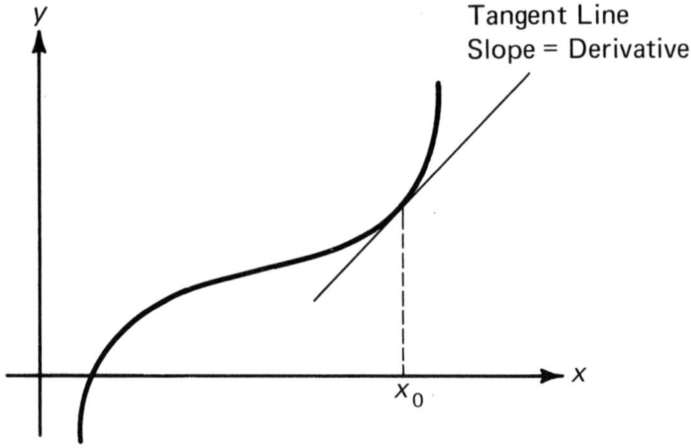

Figure 2–4. Graphic Explanation of a Derivative

change of a quantity is proportional to the amount present. A precise definition of the derivative is obtained by expressing these interpretations analytically.

The symbols $y' = f'(x)$, dy/dx, $df(x)/dx$, or $d/dx\, f(x)$ can be used to represent the derivative. It is presupposed that the reader has a working knowledge of the derivative of certain elementary functions. Some of the general formulas and rules of differentiation are given in table 2–1.

The unit step function illustrated in figure 2–2, and the function $y = 1/x$ illustrated in figure 2–3 were noted to be discontinuous at the point $x = 0$. These functions also do not possess derivatives—that is, are nondifferentiable at this point. In fact all of the previous examples of functions that had points of discontinuity were nondifferentiable at those points. However, the function $y = \sqrt{x^2}$ or $y = |x|$ (absolute value of x) shown in figure 2–5 is continuous even though there is an abrupt change in the slope of the graph. This function is not, however, differentiable at $x = 0$ (that is, the derivative is not defined at $x = 0$).

The derivative of $y = |x|$ for $x > 0$ (right-hand derivative) is $+1$, and for $x < 0$ (left-hand derivative) is -1. Whenever a function has different right-hand and left-hand derivatives at a point, the function is considered to be nondifferentiable there. Visually, the graph of the function has a corner at this point. Right-hand and left-hand derivatives of a function at a point refer to derivatives of the function as this point is approached from the right and left, respectively.

Example Problems. For the function $y = x^2$, find the derivative dy/dx. Solution:

Table 2–1
Formulas and Rules of Differentiation

Function	*Derivative*	
$y = c$	$dy/dx = 0$	(c = constant)
$y = x^a$	$dy/dx = ax^{a-1}$	($-\infty < a < \infty$; $x \neq 0$ if $a < 1$)
$y = cf$	$dy/dx = c(df/dx)$	(f = function)
$y = f + g$	$dy/dx = df/dx + dg/dx$	(g = function)
$y = fg$	$dy/dx = [(g(df/dx) - f(dg/dx)]/g^2$	
$y = f^a$	$dy/dx = af^{a-1}\, df/dx$	($f \neq 0$)

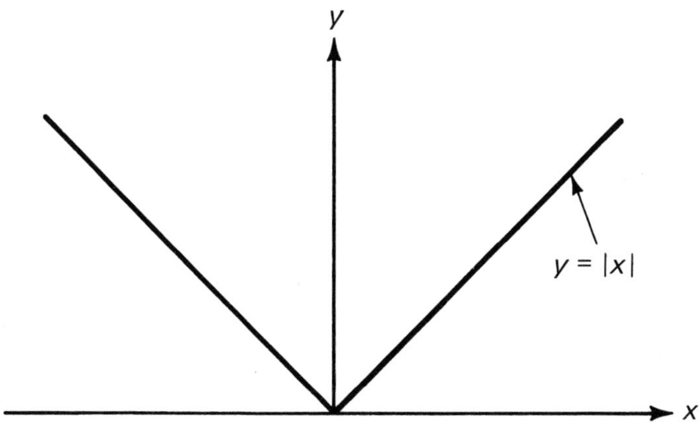

Figure 2–5. Function that is Continuous but Not Differentiable at $x = 0$

$$\frac{dy}{dx} = 2x$$

For the function $y = x^3$, find the derivative dy/dx. Solution:

$$\frac{dy}{dx} = 3x^2$$

For the function $y = 1/x$, find the derivative dy/dx. Solution:

$$\frac{dy}{dx} = -x^{-2} = -\frac{1}{x^2}$$

For the function $y = (x^2 + 2)^2 (x^2 - 2)^3$, find the derivative dy/dx. Solution:

$$\frac{dy}{dx} = (x^2 + 2)^2 \, d\frac{(x^2 - 2)^3}{dx} + (x^2 - 2)^3 \, d\frac{(x^2 + 2)^2}{dx}$$

$$d\frac{(x^2 - 2)^3}{dx} = 3(x^2 - 2)^2 \, d\frac{(x^2 - 2)}{dx} = 3(x^2 - 2)^2 \, 2x$$

$$d\frac{(x^2 + 2)^2}{dx} = 2(x^2 + 2) \, d\frac{(x^2 + 2)}{dx} = 2(x^2 + 2) \, 2x$$

$$\frac{dy}{dx} = (x^2 + 2)^2 \, 3(x^2 - 2)^2 \, 2x + (x^2 - 2)^3 \, 2(x^2 + 2) \, 2x$$

For the function $y = (x^3 + 1)/(x^4 - 2)$, find the derivative dy/dx. Solution:

$$\frac{dy}{dx} = \frac{(x^4 - 2) \, d\frac{(x^3 + 1)}{dx} - (x^3 + 1) \, d\frac{(x^4 - 2)}{dx}}{(x^4 - 2)^2}$$

$$\frac{dy}{dx} = \frac{(x^4 - 2) \, 3x^2 - (x^3 + 1) \, 4x^3}{(x^4 - 2)^2}$$

Another reason for a continuous function not to be differentiable is because the derivative may become infinite at certain points. For example, consider the function $y = \sqrt{x}$, illustrated in figure 2–6. This function is continuous everywhere, but is not differentiable at $x = 0$, since the derivative becomes infinite there. This nondifferentiability at $x = 0$ is true of all the power functions $y = x^a$, where $a < 1$.

It can be shown that if a function is differentiable at a point (some value of x), it is continuous at that point. The converse of this is not true, as illustrated by the function $y = |x|$.

Much of the terminology needed to describe the functions of interest in environmental modeling has now been introduced, and in the next section attention is focused on its application. For the convenience of the reader, we list here various terms that have been used to describe functions:

The *independent variable* to which we assign values arbitrarily

The *interval* of numbers from which these values are chosen

The *dependent variable* that is assigned, by the rule of the function, a value (*single-valued* function) or several values (*multiple-valued* function) with each assigned value of the independent variable

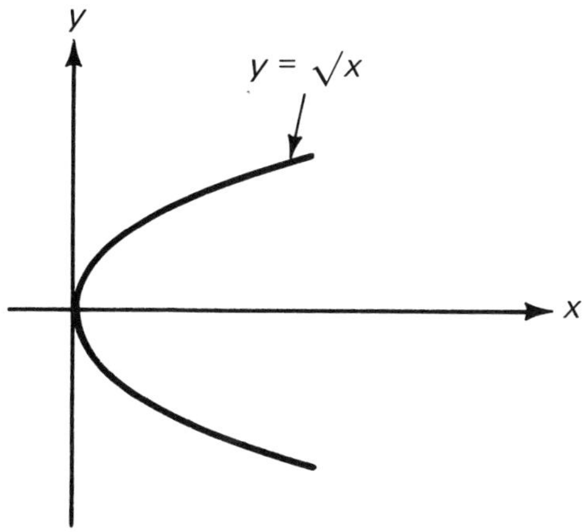

Figure 2–6. Plot of Function $y = \sqrt{x}$

The function that is *monotonic* always increases or always decreases

A function that is *continuous* has a graph which is an unbroken curve

The *derivative* of a function determines the instantaneous rate of change of the function; if a function does not possess a derivative at some point it is said to be *nondifferentiable* at that point

Types of Elementary Functions

Two basic types of elementary functions—polynomials and rational functions—are formed by using the elementary arithmetic operations of addition, subtraction, multiplication, and division applied to the variable x. Implicit in the statement that differentiability implies continuity is the fact that *polynomial* functions

$$f(x) = a + bx + cx^2 + dx^3 + \ldots.$$

are continuous for all values of x in the range $-\infty < x < \infty$. *Rational functions* (ratios of polynomials)

$$f(x) = \frac{a + bx + cx^2 + \ldots.}{A + Bx + Cx^2 + \ldots.}$$

where the numerator and the denominator are polynomials, are also continuous for all values of x except where the denominator vanishes.

As more complicated applications of arithmetic are applied, such as roots of rational functions, we are led to *algebraic functions*. Examples of algebraic functions are:

$$y = \sqrt{x}$$

$$y = \sqrt{(1 + x)}$$

$$y = \frac{1}{\sqrt{(x^2 - 1)}}$$

$$y = x + \sqrt{x}$$

The last type of functions to be treated in this text are the *transcendental functions*. Trigonometric, exponential, and logarithmic functions belong in this category. Trigonometric functions have been defined by alluding to certain characteristic dimensions associated with a circle of unit radius. Referring to figure 2–7a, we have

$$\text{OB} = \text{radius of circle} = 1$$

$$\text{OA} = \cos x$$

$$\text{AB} = \sin x$$

$$\text{CD} = \text{AB/OA} = \tan x$$

$$\text{EF} = \text{OA/AB} = \cot x$$

where x is the radian measured of the angle AOB = to the length of arc CB (for example, CBFGHC = 2π radians = 360°, CBFG = π = 180°, CBF = $\pi/2$ = 90°, etc.).

The plot of the functions $y = \sin x$ and $y = \cos x$ are shown in figure 2–7b. Therefore, it can be seen that the field of geometry generates this first kind of transcendental function. Exponential and logarithmic functions are not as easily defined and are discussed in detail in the following sections.

The Exponential Function

The function

$$y = a^x \tag{2.3}$$

where a is a positive number not equal to 1 and x is a real number, is called the exponential function. Graphs of this function for different values of a are given in figure 2–8.

The exponential function is a single-valued, continuous, monotonic function of x and assumes all real values greater than 0—that is, $0 < a^x <$

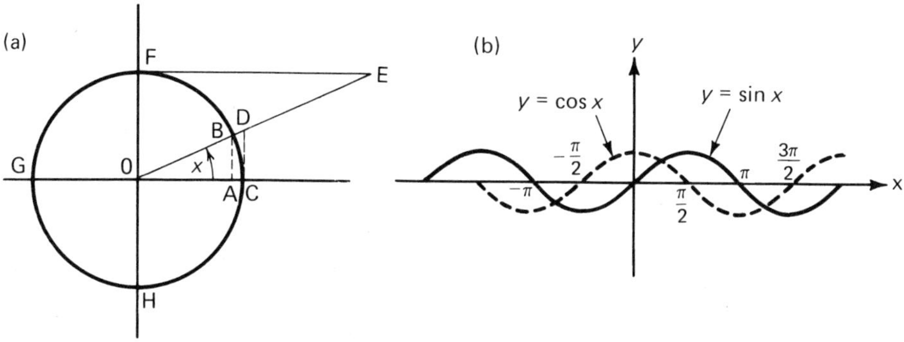

Figure 2–7. (a) Circle of Unit Radius; (b) Function $y = \sin x$, $y = \cos x$

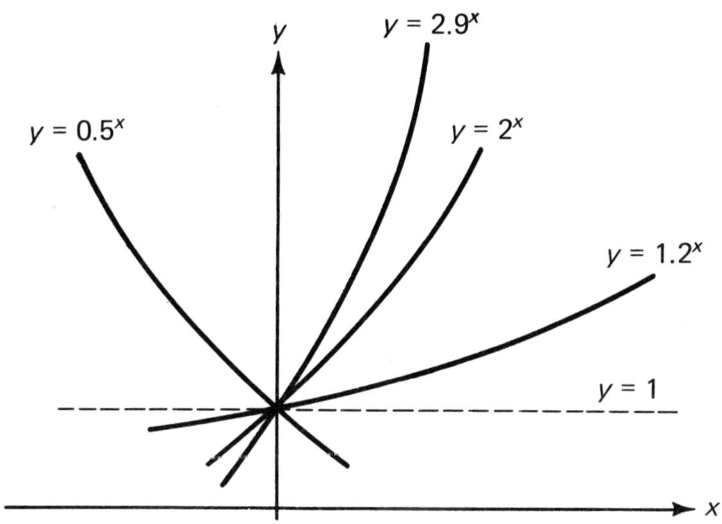

Figure 2–8. Plot of Exponential Function

∞ for $-\infty < x < \infty$. The monotonic nature of a^x is easily seen by substitution of increasing values of x; a^x always increases (monotonically increasing) for $a > 1$ and always decreases (monotonically decreasing) for $0 < a < 1$. If a computer or a scientific calculator is not readily available, then numerical values of the exponential function can be computed by the use of logarithms. Attention to the logarithm in this section is restricted to its fundamental use as a computational tool.

The logarithm of y to the base a is the exponent x in the expression $y = a^x$. This is written as

$$x = \log_a y \tag{2.4}$$

that is, the logarithm of y to the base a is that exponent x to which a must be raised to determine y. *Thus logarithms are exponents*. Two special values, easily obtained from the basic definitions for $x = 1$ and $y = 1$, respectively, are

$$\log_a a = 1, \quad \log_a 1 = 0 \tag{2.5}$$

Certain fundamental algebraic properties of exponents and logarithms are given in table 2–2.

These definitions and properties can be used to solve equations such as $2^y = 3^x$. Taking the logarithm with the base 10 ($a = 10$) of both sides of the equation yields $y \log_{10} 2 = x \log_{10} 3$, or $y = x(\log_{10} 3)/(\log_{10} 2)$. Using a table of logarithms to the base 10 yields $y = 1.5850x$, correct to five significant figures.

For an instructive example consider the following: The number of bacteria n in a certain culture at the end of t hours is given by $n = 10^{2.00+0.25t}$. The number at time $t = 0$ is $n = 10^{2.00+0.25(0)} = 10^{2.00} = 100$. After 1 hour $n = 10^{2.25}$. To determine n we take the logarithm of both sides of this equation, which yields $\log_{10} n = 2.25 \log_{10} 10$. Then n, which has a logarithm to the base 10 of 2.25 (note $\log_{10} 10 = 1$), equals 178 (to three significant figures). The procedure to find the doubling time is basically the same as that used to determine n after 1 hour. Since $n = 100$ at $t = 0$, the population is doubled when $n = 200$. Thus set $n = 200$ and solve for t:

$$\log_{10} 200 = \log_{10} 10^{2.00+0.25t}$$

or $2.30 = 2.00 + 0.25t$, which yields $t = 1.20$ hours.

The Exponential Function of Base e

There is a special value of a that because of its great importance in mathematics is worthy of special consideration. This value, which is denoted by the symbol e, is a fundamental constant like the number π. To ten significant figures,

$$e = 2.71828 \quad 1828$$

$$\log_{10} e = 0.434294 \quad 4819$$

To show why this special value of a evolves, consider the derivative of the function $y = f(x) = a^x$. First, recall from basic calculus that the slope of a straight line through the point $f(x)$ and a neighboring point $f(x+\Delta x)$ (see figure 2–9) is given by

Table 2–2
Properties of Exponents and Logarithms

$a^x a^z$	$=$	a^{x+z}
$a^x b^x$	$=$	$(ab)^x$
$(a^x)^z$	$=$	a^{xz}
$\log_a (yu)$	$=$	$\log_a y + \log_a u$
$\log_a (y/u)$	$=$	$\log_a y - \log_a u$
$\log_a y^u$	$=$	$u \log_a y$

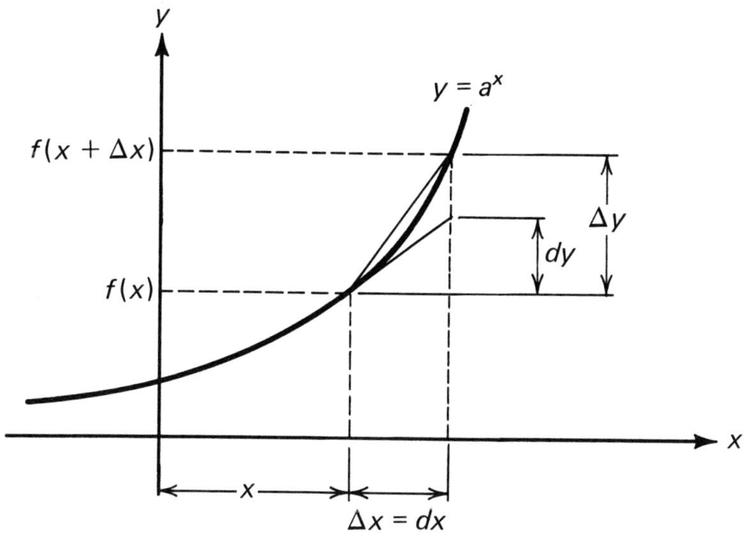

Figure 2–9. Graphic Explanation of Differential

$$\frac{\Delta y}{\Delta x} = \frac{f(x+\Delta x) - f(x)}{\Delta x} \tag{2.6}$$

and that the derivative of $f(x)$ is the slope of the tangent to the curve at the point $f(x)$. The slope of the straight line $\Delta y/\Delta x$ approaches the derivative as the neighboring point $f(x+\Delta x)$ is taken closer and closer to the point $f(x)$. Symbolically,

$$\frac{dy}{dx} = f'(x) = \lim_{\Delta x \to 0} \frac{\Delta y}{\Delta x} = \lim_{\Delta x \to 0} \frac{f(x+\Delta x) - f(x)}{\Delta x} \tag{2.7}$$

Note that in this operation only Δx varies and approaches 0. Considering the specific function $f(x) = a^x$, the slope of a straight line through the points $y = a^{(x+\Delta x)}$ and $y = a^x$ is

$$\frac{\Delta y}{\Delta x} = \frac{a^{(x+\Delta x)} - a^x}{\Delta x} = a^x\left(\frac{a^{\Delta x} - 1}{\Delta x}\right) \tag{2.8}$$

This equation illustrates that the slope of a straight line through two consecutive points on the graph of the function $y = a^x$ is proportional to the function itself. That is,

$$\frac{\Delta y}{\Delta x} = \left(\frac{a^{\Delta x} - 1}{\Delta x}\right)y = ky \tag{2.9}$$

It is easily seen that

$$\frac{dy}{dx} = \left(\lim_{\Delta x \to 0} k\right)y \tag{2.10}$$

The derivative dy/dx is therefore determined if we know the behavior of $\lim k$ as $x \to 0$. We note that the proportionality constant $k = (a^{\Delta x} - 1)/\Delta x$ approaches 0 as a approaches 1—that is, (a + 1.1, 1.01, 1.001, . . .), since the numerator $(a^x - 1)$ approaches 0 as a approaches 1 for any fixed value of x ($1^x = 1$ for any x). We also note that k always increases (monotonically increases) with a, since the numerator increases monotonically. That is,

$$k \to 0 \text{ as } a \to 1$$

$$k \to \infty \text{ as } a \to \infty$$

This monotonic behavior of k is illustrated for values of a = 1.5, 2, 2.5, 3, 3.5, and for various values of Δx in table 2–3.

Since k is monotonically increasing as a function of a from its minimum values of 0, there must be some value of a such that $k = 1$. For this value of a, $dy/dx = y$; that is, the function a^x is its own derivative. We attempt to more accurately determine a such that $k = 1$ in table 2–4. We start with $a = 2.75$ (the average of 2.5 and 3.0; see table 2–3) and use a value of Δx such that k is correct to five decimal places. The last value of a (= 2.71828) gives $k = 1$ to five decimal places. As was mentioned earlier, this value of $a = 2.71828$ is denoted by the symbol e. Because of its prominence, the function $y = e^x$ is called *the* exponential function. It is differentiated according to the formula

$$\frac{d}{dx}e^x = e^x \tag{2.11}$$

Table 2–3
Monotonic Behavior of k
(for a = 1.5, 2, 2.5, 3, and 3.5, and various Δx)

Δx \ k	$(1.5^{\Delta x} - 1)/\Delta x$	$(2^{\Delta x} - 1)/\Delta x$	$(2.5^{\Delta x} - 1)/\Delta x$	$(3^{\Delta x} - 1)/\Delta x$	$(3.5^{\Delta x} - 1)/\Delta x$
.1	0.41380	0.71773	0.95958	1.16123	1.33462
.01	0.40629	0.69555	0.92050	1.01467	1.26064
.001	0.40555	0.69339	0.91671	1.00922	1.25355
$\rightarrow 0$	0.40547	0.69315	0.91629	1.00986	1.25276

Table 2–4
Values of a with $k \cong 1$

a	2.75	2.72	2.71	2.715	2.7175	2.71875	2.71813	2.71844	2.71828
k	1.01160	1.00063	0.99695	0.99879	0.99971	1.00017	0.99994	1.00005	1.00000

Thus the function e^x (or ce^x, where c is a constant) is not changed by differentiation and can therefore be differentiated repeatedly, always reproducing itself. No other function remains unchanged by differentiation. For the purpose of this text it is the property of e^x (or ce^x) which singles it out as a function of major importance—that is, e^x is a function which changes at a rate proportional to itself and is the only function which satisfies the differential equation (in other words, a relationship between a function and its derivatives)

$$\frac{dy}{dx} = y \tag{2.12}$$

This property is basic to many biological and chemical processes and is sometimes called the *law of natural growth*.

The exponential function usually occurs in the form e^{kx} in physical problems. By a change of scale—that is, multiplication of x by a constant k—the function $y = e^{kx}$ is obtained from $y = e^x$, and its derivative is ke^{kx}. That is

$$\frac{dy}{dx} = \frac{d}{dx} e^{kx} = ke^{kx} = ky \tag{2.13}$$

Plots of e^{kx} for various values of k are shown in figure 2–10. It is easily seen from figure 2–10 that changing the sign of k is equivalent to rotating the curve 180° about the y axis. Thus the graphs of the functions e^{kx} with k alternately positive and negative are mirror images of each other about the y axis.

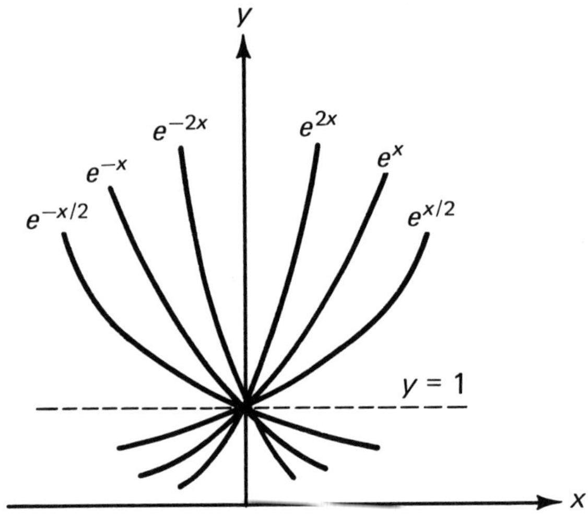

Figure 2–10. Plot of $y = e^{kx}$ for Variable k

Differentiation of the general exponential function $y = a^x$ through the use of the logarithmic function is discussed in the next section. It is possible, however, to differentiate $y = a^x$ according to the following equation

$$\frac{d}{dx} a^x = \frac{d}{dx} e^{kx} = ke^{kx} = ka^x \tag{2.14}$$

Since $e^{kx} = (e^k)^x$ by one of the fundamental properties of exponents (table 2–2), if $a^x = e^{kx}$, then $e^k = a$. Taking the logarithm to the base 10 of both sides of this relationship yields $k \log_{10} e = \log_{10} a$, and therefore $k = \log_{10} a / \log_{10} e$. The value of the denominator to ten places has already been given, and it only remains to calculate $\log_{10} a$ for any particular value of a. Thus it is possible to calculate k and consequently the derivative defined by equation (2.14).

The development from equation (2.11) through equation (2.14) underscores the most important concept concerning the exponential functions: The only function that satisfies the relationship

$$\frac{dy}{dx} = ky$$

is the special exponential function $y = e^{kx}$ (or $y = a^x$, where $a = e^k$), which states that the rate of change of the function y is proportional to itself.

Inverse Function, Logarithmic Function

Solving for x in terms of y in the function

$$y = f(x) \tag{2.15}$$

yields the inverse function

$$x = f^{-1}(y) = G(y) \tag{2.16}$$

when this inversion is possible. In previous discussions it was pointed out that some functions have multiple-valued inverses. Thus, for example, the function $y = x^2$ has an associated multiple-valued inverse function $x = \pm \sqrt{y}$. Only single-valued inverse functions will be treated in this text unless noted otherwise. Table 2–5 shows certain common functions and their corresponding inverse functions. The last two examples in table 2–5 show multiple-valued inverse functions.

Note that the functions $y = f(x)$ and $x = f^{-1}(y)$ are equivalent, and if plotted on the same rectangular coordinate system they would yield identical graphs. If we interchange the roles of x and y in the inverse function $x = f^{-1}(y)$, the function $y = f^{-1}(x) = G(x)$ [see equation (2.16)] is obtained. In this form we can study the inverse function with x and y in the usual roles of independent and dependent variable, respectively. The graph of the function $y = f^{-1}(x)$ will be a mirror image of the graph of $y = f(x)$ in the line $y = x$. This property is illustrated for the functions $y = x^2$ and $y = e^x$ in figure 2–11.

The exponential function with the base a,

$$y = a^x \tag{2.17}$$

and the logarithmic function to the base a,

$$x = \log_a y \tag{2.18}$$

were defined previously [see equations (2.3) and (2.4)]. Since these functions are in the form of equations (2.15) and (2.16), $x = \log_a y$ is defined as the inverse function of $y = a^x$. When $a = e$, equations (2.17) and (2.18) become

$$y = e^x \tag{2.19}$$

and

$$x = \log_e y \tag{2.20}$$

the logarithm to the base e of x is called the natural logarithm of x, and will be written as $x = \ln y$—that is,

$$\log_e y = \ln y \tag{2.21}$$

Table 2–5
Inverse Functions

$y = cx$	$x = (1/c)\,y$
$y = cx^3$	$x = (1/c)^3\sqrt{y}$
$y = cx^a$	$x = \sqrt[a]{y/c}\ (a = 1, 3, 5, \ldots)$
$y = \sin x$	$x = \sin^{-1} y$
$y = \tan x$	$x = \tan^{-1} y$

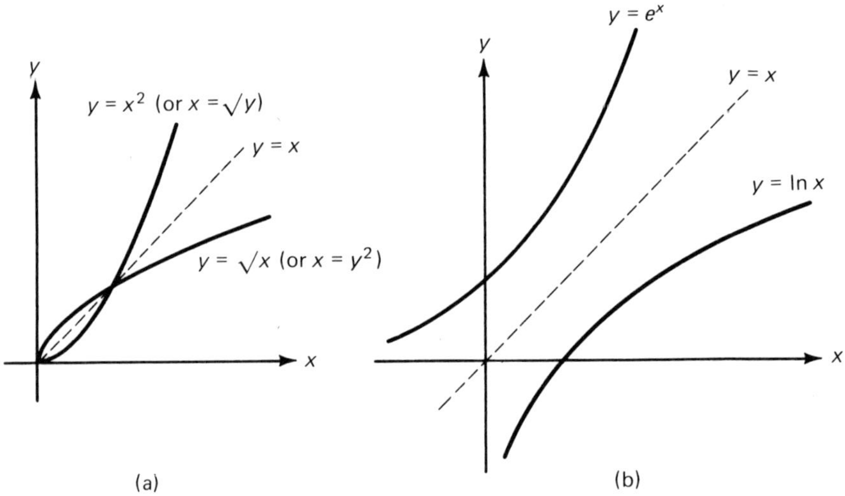

Figure 2–11. Plot of Reciprocal Functions

The special relations

$$\ln e^x = x \tag{2.22}$$

and

$$e^{\ln y} = y \tag{2.23}$$

are easily obtained from equations (2.19) and (2.20), and the definition (2.21).

The logarithmic function $y = \ln x$ [where we have interchanged the roles of x and y from the previously defined logarithmic function, equation (2.21)] is of primary importance in applied mathematics, and is defined as the function

$$y = \ln x = \int_1^x \frac{dt}{t} \tag{2.24}$$

for $x > 0$. The derivative of the logarithmic function of equation (2.24) can be obtained as

$$y = \frac{d}{dx}(\ln x) = \frac{1}{x} \tag{2.25}$$

from equation (2.24) by an application of the fundamental theorem of calculus.

The models of many environmental processes utilize the general exponential and logarithmic functions. Applied mathematics, however, deal primarily with the special exponential function $y = e^x$ and the natural logarithmic function $y = \ln x$. We therefore next develop relationships between these two sets of functions so that the techniques of applied mathematics may be used.

Applying the natural logarithm to both sides of $y = a^x$ yields $\ln y = x \ln a$ or $x = (\ln y)/(\ln a)$. Since $x = \log_a y$ (when $y = a^x$), setting these two expressions for x equal to each other yields

$$\log_a y = \frac{\ln y}{\ln a} \tag{2.26}$$

Substituting $y = a^x$ into both sides of equation (2.23) and applying the last of the fundamental properties of logarithms (table 2–2) yields

$$a^x = e^{\ln a^x} = e^{x \ln a} \tag{2.27}$$

We now use the relationships $y = e^x$ and $y = \ln x$ to differentiate the general exponential function $y = a^x$ and the general logarithmic function $y = \log_a x$. Thus equations (2.17) and (2.18) can be rewritten as

$$y = e^{x \ln a} \tag{2.28}$$

and

$$x = \frac{\ln y}{\ln a} \tag{2.29}$$

The derivatives of the functions $y = a^x$ and $y = \log_a x$ can now be determined as

$$\frac{d}{dx} a^x = \frac{d}{dx} e^{x \ln a} = e^{x \ln a} \ln a = a^x \ln a \tag{2.30}$$

and

$$\frac{d}{dx} \log_a x = \frac{d}{dx} \frac{\ln x}{\ln a} = \frac{1}{x \ln a} \tag{2.31}$$

When $a = e$, these relationships reduce to the simple forms given in equations (2.11) and (2.25).

3 Mathematics of Material Transport

Material transport is the movement (flux) of material (biological or chemical species) from one location to another. The transport of any material in a fluid occurs through the processes of advection and/or diffusion. Advection is material transport resulting from fluid movement, while diffusion refers to material movement in solution due to differences in concentration throughout the fluid. We now consider in detail the relationships governing material transport in fluid systems.

Mass Transport Rates—The Partial Derivative

Natural forces tend to distribute substances uniformly throughout a fluid by material transport. A fluid that has a uniform composition at all points within its boundary is called *homogeneous*. If substances within the fluid are not distributed uniformly, then the concentrations of its components vary throughout, and concentration gradients (rate of change of concentration with position) exist. In this section we describe mathematically the forces that minimize these concentration gradients. The following discussion does not consider material transport due to pressure gradients, thermal gradients, or the effect of unequal forces acting on different components.

The concentration of material A, C_A (moles l^{-1}), in a fluid is in general dependent on (a function of) both position (spatial rectangular coordinates x, y, z located with respect to a *fixed* reference frame) and time (temporal coordinate t). Thus we can write

$$C_A = f(x,y,z,t) \tag{3.1}$$

To simplify the following discussion we consider only functions dependent on the x coordinate. This results in no loss of generality, since the resulting equations are easily generalized to include the other spatial coordinates, y and z, due to symmetry considerations. We therefore write equation (3.1) in the form

$$C_A = f(x,t) \tag{3.2}$$

The concentration gradient of material A can now be defined as $\partial C_A / \partial x$, where the "curly d" notation is used to denote the partial derivative of C_A with respect to x. This derivative is determined by holding t constant and treating C_A as if it was a function of x alone. Similarly, the partial derivative of C_A with respect to t—that is, $\partial C_A/\partial t$—is obtained by holding x constant and treating C_A as a function of t alone.

To illustrate the treatment of partial derivatives, consider the following examples:

Example 3.1. Given the relationship for the concentration of A, as a function of time and distance

$$C_A = C_0 x^{-1} e^{-kt}$$

where C_0 = initial concentration of A and k = specific rate constant, the concentration gradient is given by

$$\frac{\partial C_A}{\partial x} = -C_0 x^{-2} e^{-kt}$$

and the partial derivative with respect to t is

$$\frac{\partial C_A}{\partial t} = -kC_0 x^{-1} e^{-kt}$$

In a similar manner, the concentration gradient and partial derivative with respect to t are illustrated in examples 3.2 and 3.3 for different concentration functions:

Example 3.2.

$$C_A = C_0 x(x + t)^{-1}$$

$$\frac{\partial C_A}{\partial x} = C_0 t(x + t)^{-2}$$

$$\frac{\partial C_A}{\partial t} = -C_0 x(x + t)^{-2}$$

Example 3.3.

$$C_A = C_0 e^{-kx}(\ln 1 - \ln t)$$

$$\frac{\partial C_A}{\partial x} = -C_0 k e^{-kx}(\ln 1 - \ln t)$$

$$\frac{\partial C_A}{\partial t} = C_0 e^{-kx} \ln t^{-1}$$

These examples reveal that the partial derivative of a function of two independent variables is formed according to the same rules as the derivative of a function of one independent variable by considering the other variable as a constant.

Physically, the gradient $\partial C_A / \partial x$ represents the rate of change of C_A in the x direction, holding time fixed, and $\partial C_A / \partial t$ represents the rate of change of C_A that occurs within a stationary volume element, at a fixed value of x, through which the fluid is flowing. The physical meaning of the above partial derivatives and other rates of change of C_A may be better illustrated by the following example:

We have an instrument that measures C_A continuously at a fixed location in a flowing stream—for example, at the rear of an anchored boat. The difference in successive values of C_A is $\partial C_A / \partial t$ and is called the *local rate of change* of C_A. If another nearby instrument was simultaneously measuring C_A during the same time periods, the difference between the two instrument readings divided by the distance (say dx) between the instruments would be represented by the gradient $\partial C_A / \partial x$. Note that keeping x fixed means that successive readings are taken at the same location, and keeping t fixed means that readings of two nearby locations are taken simultaneously.

Environmental models often require the use of another type of time derivative. This is called the *total* time derivative dC_A / dt and reflects the change in C_A due to local changes, gradients, and instrument velocity. Referring back to the example of measuring C_A continuously from the boat in the stream, it can be seen that the *total* time derivative represents all the rates of change occurring in C_A even if the boat is moving at some velocity while recording concentrations of C_A. Mathematically, the total time derivative can be defined by applying the "chain rule," or through the use of the Taylor series expansion. Because of its relative completeness and clarity in presentation, we develop the total time derivative utilizing the Taylor series.

Taylor Series Expansion—The Total Time Derivative

When mathematically modeling physical systems, it is often necessary to determine a difference in the value of some quantity (for example, concentration, velocity, etc.) at either of two neighboring points (that is, x and $x + \Delta x$) or at different times (that is, t and $t + \Delta t$), or both. Taylor's series is a tool that can be used to approximate these differences.

Suppose, for example, we know the value of the concentration of material A, C_A, at some position x and time [that is, we know $C_A(x,t)$], and also that we know the value of all the derivatives of C_A at this point. We denote the difference in the values of $C_A(x,t)$ and $C_A(x + \Delta x, t + \Delta t)$ as ΔC_A. That is,

$$\Delta C_A = C_A(x + \Delta x, t + \Delta t) - C_A(x,t)$$

Then, by Taylor's series we can write this difference ΔC_A as

$$\Delta C_A = \frac{\partial C_A}{\partial x} \Delta x + \frac{\partial C_A}{\partial t} \Delta t + \frac{1}{2} \frac{\partial^2 C_A}{\partial x^2} (\Delta x)^2 + \frac{\partial^2 C_A}{\partial x \partial t} \Delta x \Delta t + \frac{\partial^2 C_A}{\partial t^2} (\Delta t)^2 + \ldots . \tag{3.3}$$

Equation (3.3) does not require the values of the increments x and t to be "small"; but, for our purposes, we consider a special case of this equation as these increments become vanishingly small (Δx, $\Delta t \rightarrow 0$). In this case equation (3.3) can be written as

$$dC_A = \frac{\partial C_A}{\partial x} dx + \frac{\partial C_A}{\partial t} dt \tag{3.4}$$

In equation (3.4) we have used the notation dC_A to represent the linear portion of ΔC_A, denoted Δx and Δt as dx and dt ($\Delta x \equiv dx$, $\Delta t \equiv dt$), and have neglected the higher order terms $(\Delta x)^2$, $\Delta x \Delta t$, $(\Delta t)^2$, etc. which become negligible as Δx, $\Delta t \rightarrow 0$.

Dividing equation (3.4) by dt gives the total change in C_A, (dC_A), during the time interval dt:

$$\frac{dC_A}{dt} = \frac{\partial C_A}{\partial x} \frac{dx}{dt} + \frac{\partial C_A}{\partial t} \tag{3.5}$$

The term dx/dt is the rate at which the distance between the neighboring points is traveled—that is, the velocity of the boat in our example. Equation (3.5) is an expression which also illustrates the results of applying the "chain rule" to a function of the two variables x and t.

We have now defined [by equation (3.5)] the total time derivative as it evolves from the Taylor series. We are, however, most interested in a special type of total time derivative which reflects a change in C_A in an element of fluid as it travels with the stream flow toward or away from the reference point. Again referring to our boat example, we see that our boat must monitor C_A while moving at the same velocity as the stream (following the same element of fluid) to obtain this special total derivative. This rate of change has received special attention in the literature and is usually denoted by the symbol DC_A/Dt. This special derivative has been given various names such as material, convective, substantial, particle, comoving, and the derivative following the motion. Thus

$$\frac{DC_A}{Dt} = \frac{\partial C_A}{\partial x} V_x + \frac{\partial C_A}{\partial t} \tag{3.6}$$

This is seen to be a special case of equation (3.5) where we have set dx/dt (the speed of the boat) equal to V_x (the speed of the element of liquid in the flow).

In three dimensions this equation takes the form

$$\frac{DC_A}{Dt} = \left(\frac{\partial C_A}{\partial x} V_x + \frac{\partial C_A}{\partial y} V_y + \frac{\partial C_A}{\partial z} V_z\right) + \frac{\partial C_A}{\partial t} \tag{3.7}$$

We now consider another example to illustrate the mechanics of the additional derivatives introduced earlier:

Example 3.4. The concentration of an element A is given by the relationship

$$C_A = C_0 x^{-1} e^{-kt}$$

It is given that $dx/dt = \alpha V_0$, where α = constant and $V_x = V_0$ = velocity of the fluid and assumed constant. Note that $dx/dt \neq V_0$. Then the local rate of change is

$$\frac{\partial C_A}{\partial t} = -kC_0 e^{-kt} x^{-1}$$

and the gradient is given by

$$\frac{\partial C_A}{\partial x} = -C_0 e^{-kt} x^{-2}$$

From equation (3.5) we can now develop the total time derivative as

$$\frac{dC_A}{dt} = -kC_0 e^{-kt} x^{-1} + (-C_0 e^{-kt} x^{-2})(\alpha V_0)$$

Simplifying, we obtain

$$\frac{dC_A}{dt} = -C_0 x^{-1} e^{-kt}(k + \alpha V_0 x^{-1})$$

In our special case—that is, $V_x = V_0$, DC_A/Dt may be determined from equation (3.6) as

$$\frac{DC_A}{Dt} = (-kC_0 x^{-1} e^{-kt}) + (-C_0 x^{-2} e^{-kt}) V_0$$

or simplifying, as

$$\frac{DC_A}{Dt} = -C_0 x^{-1} e^{-kt}(k + x^{-1}V_0)$$

Note the following special cases:

$\alpha = 0$ (boat velocity = 0, anchored boat): $\dfrac{dC_A}{dt} = \dfrac{\partial C_A}{\partial t}$

$\alpha = 1$ (boat velocity = velocity of fluid): $\dfrac{dC_A}{dt} = \dfrac{DC_A}{Dt}$

$V_0 = 0$ (quiescent system): $\dfrac{DC_A}{Dt} = \dfrac{\partial C_A}{\partial t}$

We now return to our discussion of material transport by identifying the two types of transport defined by equation (3.6). In a quiescent system—that is, $V_x = 0$—equation (3.6) reduces to $DC_A/Dt = \partial C_A/\partial t$. That is, the total change in C_A with time is that which occurs within a stationary volume element at a fixed x coordinate. Since $V_x = 0$—that is, no fluid flowing through the volume element—the change in C_A is then caused solely by molecular action and is called *diffusive* or *dispersive* mass transport. This is also the type of material transport that an observer would see moving along with the fluid element in a nonquiescent system. (Note that in both cases the observer always sees the same element of fluid.) It now can be seen that the appearance of the term $(\partial C_A/\partial x)V_x$ in equation (3.6) is due to the fact that the observer is using a fixed coordinate system to reference his position as he moves with the flow. If the observer uses a reference frame moving with the same velocity as the fluid, then we would again have $DC_A/Dt = \partial C_A/\partial t$. Using a moving reference frame would appear to be a simpler approach; however, the mathematical complexity that would be encountered in the solution of models using this approach precludes it. The additional term $(\partial C_A/\partial x)V_x$ in equation (3.6), which is due to the use of a fixed reference frame, is called *advective* or *convective* material transport.

We now recall the different types of derivatives for the reader to review:

$\partial C_A/\partial t$ is the derivative that represents the rate of change of C_A within a stationary volume element through which fluid is flowing. In this case the observer is located at a fixed location.

$\partial C_A/\partial x$ is the derivative that represents the rate of change in C_A between neighboring elements of fluid at any instant of time. Determination of this derivative requires two different observers.

dC_A/dt is the derivative that represents the rate of change of C_A as the observer moves throughout the fluid, but not at the same velocity as the flow. This is a *total* time derivative.

DC_A/Dt is the derivative that represents the change in C_A within a *particular element of fluid* as it moves with the flow. In this case the observer moves with the flow, and it too is a *total* time derivative.

Equation of Continuity

In this section we apply the law of conservation of mass to a material A that is flowing with the fluid through a stationary volume element $\Delta x \Delta y \Delta z$. To best describe the various rate processes involved, we derive the continuity equation in terms of the mass concentration of material A—that is, ρ_A (mgl^{-1}) instead of the molar concentration C_A (moles l^{-1}). Other terms that will be used in this derivation are the velocity of material A in the x direction V_{Ax} (cm sec^{-1}), and the rate of chemical and/or biological mass production of material A, r_A (mgl^{-1} sec^{-1}).

The time rate of change in mass of material A in the volume element (see figure 3–1) is

$$\frac{\partial \rho_A}{\partial t} \Delta x \Delta y \Delta z \tag{3.8}$$

and the rate of mass input (flux) into the volume element at x (we continue our development in terms of x only) is

$$\rho_A V_{Ax} \Delta y \Delta z \tag{3.9}$$

The rate of mass output (flux) from the volume element at $x + \Delta x$, where $\Delta(\rho_A V_{Ax})$ is the difference in the value of the flux at the points x and $x + \Delta x$, is

$$[\rho_A V_{Ax} + \Delta(\rho_A V_{Ax})] \Delta y \Delta z \tag{3.10}$$

and the rate of chemical/biological production in the volume element by reaction is

$$r_A \Delta x \Delta y \Delta z \tag{3.11}$$

The symbol n_{Ax} ($= \rho_A v_{Ax}$) is often used to denote mass flux of material A. The material A must be conserved, thus

rate of change of
A in volume
element = [input flux] − [output flux] + [reactions]

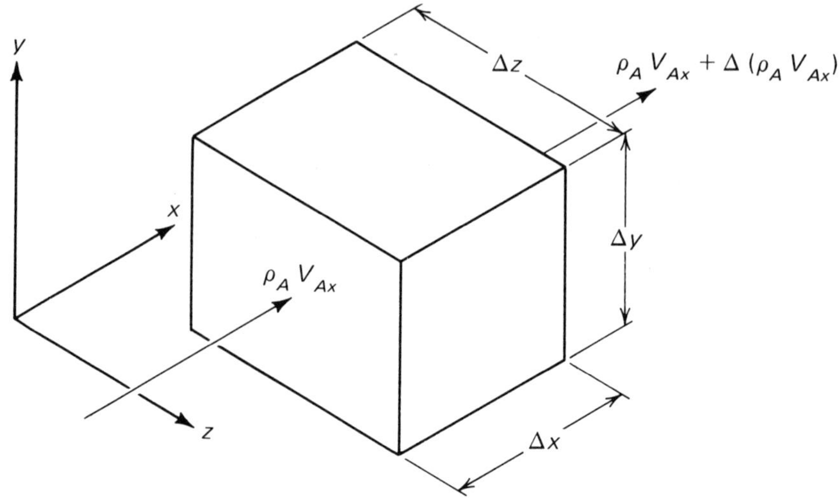

Figure 3–1. Continuity Volume Element

or

$$\frac{\partial \rho_A}{\partial t} \Delta x \Delta y \Delta z = \rho_A V_{Ax} \Delta y \Delta z - [\rho_A V_{Ax} + \Delta(\rho_A V_{Ax})] \Delta y \Delta z + r_A \Delta x \Delta y \Delta z \tag{3.12}$$

Dividing by $\Delta x \Delta y \Delta z$ and taking the limit as these quantities approach 0 yields

$$\frac{\partial \rho_A}{\partial t} = -\frac{\partial(\rho_A V_{Ax})}{\partial x} + r_A \tag{3.13}$$

We can now express the continuity equation in terms of molar concentration C_A without introducing a new velocity term—that is, molar average velocity. It is important for the reader at this point to realize that equation (3.13) was not derived using molar concentrations, because moles are not necessarily conserved while mass is. It can be seen from the disassociation of one molecule of water

$$H_2O \rightarrow H^+ + OH^-$$

that while 18 g of water produces 1 g of hydrogen ions and 17 g of hydroxyl ions, 1 mole of water produces 1 mole of $[H^+]$ and 1 mole of $[OH^{-1}]$. To express the continuity equation in terms of molar concentration C_A, we divide equation (3.13) by the molecular weight of A, M_A, and get

$$\frac{\partial C_A}{\partial t} = -\frac{\partial (C_A V_{Ax})}{\partial x} + R_A \tag{3.14}$$

where $C_A = \rho_A/M_A$, and $R_A = r_A/M_A$ is the molar production of A (moles l^{-1} sec^{-1}) in the volume element.

Generalizing to three dimensions, equations (3.13) and (3.14) become

$$\frac{\partial \rho_A}{\partial t} = -\left[\frac{\partial (\rho_A V_{Ax})}{\partial x} + \frac{\partial (\rho_A V_{Ay})}{\partial y} + \frac{\partial (\rho_A V_{Az})}{\partial z}\right] + r_A \tag{3.15}$$

and

$$\frac{\partial C_A}{\partial t} = -\left[\frac{\partial (C_A V_{Ax})}{\partial x} + \frac{\partial (C_A V_{Ay})}{\partial y} + \frac{\partial (C_A V_{Az})}{\partial z}\right] + R_A \tag{3.16}$$

The reader should now review the above derivation and note that equations (3.15) and (3.16) represent the net gain of material A (mass influx) within the volume element per unit time.

Equations (3.13) through (3.16) are equations of continuity for a single component—that is, material A—in a multicomponent system in various forms. We now look at the entire system and see how continuity may be applied.

For simplicity we again return to the form given by equation (3.13). If we add like equations of continuity for all the materials in the fluid system, we would have

$$\frac{\partial \rho}{\partial t} = -\frac{\partial}{\partial x}(\rho V_x) \tag{3.17}$$

where

$$\begin{aligned} \rho &= \text{the sum of all the mass concentrations} \\ &= (\rho_A + \rho_B + \ldots.) \\ V_x &= \text{mass average velocity in the } x \text{ direction} \\ &= \rho^{-1}(\rho_A V_{Ax} + \rho_B V_{Bx} + \ldots.) \end{aligned}$$

Note that the sum of the mass production rates $(r_A + r_B + \ldots.)$ disappear due to the conservation of mass of the reacting components. Also, the mass average velocity V_x is the *same* as the velocity of the fluid as used in equation (3.6).

It is instructive to now look at some special cases of the continuity equation written on the entire fluid system. For example, for a constant mass density ρ, $\partial\rho/\partial t = 0$, equation (3.17) reduces to

$$\frac{\partial V_x}{\partial x} = 0 \tag{3.18}$$

In three dimensions, equations (3.17) and (3.18) become

$$\frac{\partial \rho}{\partial t} = -\left[\frac{\partial(\rho V_x)}{\partial x} + \frac{\partial(\rho V_y)}{\partial y} + \frac{\partial(\rho V_z)}{\partial z}\right] \tag{3.19}$$

and

$$\frac{\partial V_x}{\partial x} + \frac{\partial V_y}{\partial y} + \frac{\partial V_z}{\partial z} = 0 \tag{3.20}$$

Equations (3.19) and (3.20) reflect the fact that the total mass of the entire fluid system is also conserved.

In the case where steady state conditions prevail (no changes occurring with time), $\partial\rho_A/\partial t = 0$, equation (3.13) reduces to

$$\frac{\partial(\rho_A V_{Ax})}{\partial x} = r_A \tag{3.21}$$

and similarly from equation (3.14),

$$\frac{\partial(C_A V_{Ax})}{\partial x} = R_A \tag{3.22}$$

Equations (3.21) and (3.22) simply state that changes in the concentration of material A in the stationary volume element occur only with position and that none occur with time. In this case the gradients (spatial rate of change) of the mass and molar concentrations are directly proportional to the mass and molar rates of production, respectively. If there is no mass production within the volume element ($r_A = R_A = 0$), then the concentration gradients vanish (flux in = flux out).

In three dimensions, equations (3.21) and (3.22) become

$$\frac{\partial(\rho_A V_{Ax})}{\partial x} + \frac{\partial(\rho_A V_{Ay})}{\partial y} + \frac{\partial(\rho_A V_{Az})}{\partial z} = r_A \tag{3.23}$$

and

$$\frac{\partial(C_A V_{Ax})}{\partial x} + \frac{\partial(C_A V_{Ay})}{\partial y} + \frac{\partial(C_A V_{Az})}{\partial z} = R_A \tag{3.24}$$

Diffusive Mass Transport

Equation (3.13), which is the continuity equation for a single component A, is not in a sufficiently useful form due to the presence of a flux term $\rho_A V_{Ax}$. However, this term can be replaced by a form of Fick's first law of diffusion with respect to a fixed reference frame. [The student may have seen the more common form that excludes the $\rho_A V_x$ term, which is valid for the special case of a quiescent system ($V_x = 0$).]

$$\rho_A V_{Ax} = -\rho \mathcal{D}_{AM} \frac{\partial}{\partial x}\left(\frac{\rho_A}{\rho}\right) + \rho_A V_x \tag{3.25}$$

where $\mathcal{D}_{AM}$ is the mass diffusivity (cm^2 sec^{-1}) of material A in the mixture ($\rho_A + \rho_B + \ldots .$). Equations for the y and z direction equivalent to equation (3.25) are

$$\rho_A V_{Ay} = -\rho \mathcal{D}_{AM} \frac{\partial}{\partial y}\left(\frac{\rho_A}{\rho}\right) + \rho_A V_y \tag{3.26}$$

and

$$\rho_A V_{Az} = -\rho \mathcal{D}_{AM} \frac{\partial}{\partial z}\left(\frac{\rho_A}{\rho}\right) + \rho_A V_z \tag{3.27}$$

If we were to use a moving coordinate system, V_x would be 0, and it is seen that the flux of material A, ($\rho_A V_{Ax}$) relative to the mixture (diffusion of A) is in the direction of the decreasing mass concentration ratio. This is the diffusive transport of material A in response to a concentration gradient. For a fixed coordinate system the term $\rho_A V_x$ in equation (3.25) represents the movement of material A along with the fluid and is a convective mass transport term. This convective term would also go to 0 in a quiescent system ($V_x = V_y = V_x = 0$), so that the transport of material A is again solely in response to a concentration gradient.

Substituting Fick's law [equation (3.25) into equation (3.13)] yields

$$\frac{\partial \rho_A}{\partial t} = \frac{\partial}{\partial x}\left[\rho \mathcal{D}_{AM} \frac{\partial}{\partial x}\left(\frac{\rho_A}{\rho}\right)\right] - \frac{\partial}{\partial x}(\rho_A V_x) + r_A \tag{3.28}$$

Simplifying and collecting terms we get

$$\frac{\partial \rho_A}{\partial t} + V_x \frac{\partial \rho_A}{\partial x} + \rho_A \frac{\partial V_x}{\partial x} = \frac{\partial}{\partial x}\left[\rho \mathcal{D}_{AM} \frac{\partial}{\partial x}\left(\frac{\rho_A}{\rho}\right)\right] + r_A \tag{3.29}$$

Equations (3.28) and (3.29) are general forms of the continuity equation for material A and are valid for variable ρ (or C) and $\mathcal{D}_{AM}$. Equation (3.29) can be generalized to three dimensions as

$$\frac{\partial \rho_A}{\partial t} + V_x \frac{\partial \rho_A}{\partial x} + V_y \frac{\partial \rho_A}{\partial y} + V_z \frac{\partial \rho_A}{\partial z} + \rho_A\left(\frac{\partial V_x}{\partial x} + \frac{\partial V_y}{\partial y} + \frac{\partial V_z}{\partial z}\right) =$$
$$\frac{\partial}{\partial x}\left[\rho \mathcal{D}_{AM} \frac{\partial}{\partial x}\left(\frac{\rho_A}{\rho}\right)\right] + \frac{\partial}{\partial y}\left[\rho \mathcal{D}_{AM} \frac{\partial}{\partial y}\left(\frac{\rho_A}{\rho}\right)\right]$$
$$+ \frac{\partial}{\partial z}\left[\rho \mathcal{D}_{AM} \frac{\partial}{\partial z}\left(\frac{\rho_A}{\rho}\right)\right] + r_A \tag{3.30}$$

For most situations, the total density and diffusivity (ρ and $\mathscr{D}_{AM}$) remain constant; in this case equation (3.29) reduces to

$$\frac{\partial \rho_A}{\partial t} + V_x \frac{\partial \rho_A}{\partial x} = \mathscr{D}_{AM} \frac{\partial^2 \rho_A}{\partial x^2} + r_A \tag{3.31}$$

Similarly, from equation (3.30),

$$\frac{\partial \rho_A}{\partial t} + V_x \frac{\partial \rho_A}{\partial x} + V_y \frac{\partial \rho_A}{\partial y} + V_z \frac{\partial \rho_A}{\partial z} = \mathscr{D}_{AM} \nabla^2 \rho_A + r_A \tag{3.32}$$

where ∇^2 (read "del squared") is simply an abbreviation for

$$\frac{\partial^2}{\partial x^2} + \frac{\partial^2}{\partial y^2} + \frac{\partial^2}{\partial z^2} \tag{3.33}$$

that is, $\nabla^2 f$ is merely the sum of the second derivatives with respect to x, y, and z of a function f. We have made use here of equation (3.18),

$$\frac{\partial V_x}{\partial x} = 0$$

and equation (3.20),

$$\frac{\partial V_x}{\partial x} + \frac{\partial V_y}{\partial y} + \frac{\partial V_z}{\partial z} = 0$$

which are special cases of the continuity equation for ρ = constant. We can now divide equations (3.31) and (3.32) by M_A, ($C_A = \rho_A/M_A$), and obtain

$$\frac{\partial C_A}{\partial t} + V_x \frac{\partial C_A}{\partial x} = \mathscr{D}_{AM} \frac{\partial^2 C_A}{\partial x^2} + R_A \tag{3.34}$$

and

$$\frac{\partial C_A}{\partial t} + V_x \frac{\partial C_A}{\partial x} + V_y \frac{\partial C_A}{\partial y} + V_z \frac{\partial C_A}{\partial z} = \mathscr{D}_{AM} \nabla^2 C_A + R_A \tag{3.35}$$

It is now possible to substitute equation (3.34) into equation (3.6) to determine the total time derivative DC_A/Dt. Thus we have obtained an expression for the time rate of change in concentration of material A that an observer moving with the fluid would see, utilizing Fick's first law of diffusion and continuity equations for a material moving through a stationary volume element—that is,

$$\frac{DC_A}{Dt} = \mathscr{D}_{AM} \frac{\partial^2 C_A}{\partial x^2} + R_A \tag{3.36}$$

Similarly, substituting equation (3.35) into equation (3.7) we obtain

$$\frac{DC_A}{Dt} = \mathscr{D}_{AM}\nabla^2 C_A + R_A \tag{3.37}$$

In a *quiescent system*, with no rate of mass production (that is, $R_A = 0$), equations (3.34) and (3.35) reduce to

$$\frac{\partial C_A}{\partial t} = \mathscr{D}_{AM}\frac{\partial^2 C_A}{\partial x^2} \tag{3.38}$$

and

$$\frac{\partial C_A}{\partial t} = \mathscr{D}_{AM}\nabla^2 C_A \tag{3.39}$$

Equations (3.38) and (3.39) represent one- and three-dimensional forms of Fick's second law of diffusion.

For steady-state conditions, equations (3.34) and (3.35) reduce to

$$V_x\frac{\partial C_A}{\partial x} = \mathscr{D}_{AM}\frac{\partial^2 C_A}{\partial x^2} + R_A \tag{3.40}$$

and

$$V_x\frac{\partial C_A}{\partial x} + V_y\frac{\partial C_A}{\partial y} + V_z\frac{\partial C_A}{\partial z} = \mathscr{D}_{AM}\nabla^2 C_A + R_A \tag{3.41}$$

4 Models of Environmental Processes

The student has now reviewed the primary mathematical functions encountered in modeling common environmental processes. This chapter will introduce several basic models that are in common use by environmental engineers and scientists today. The underlying concepts for the creation of each model are the same, and the same types of relationships appear consistently throughout the models. Each model is represented by a differential equation, and the formal solutions of these equations are explained in chapter 5.

The Modeling Process

Modeling any environmental system is usually an attempt to quantify a complex biological and/or chemical system. In an attempt to obtain a workable mathematical model, many processes have to be idealized due to the lack of quantitative information on natural processes. Most models include rate terms—that is, the change in some parameter as a function of time—and attempt to predict the change of some substance in a defined volume.

The underlying principle in most models is the conservation of mass. This simply means that the mass of the substances being monitored can be accounted for by material transfer in and out of the considered volume. In addition the creation or destruction of substance due to biological and/or chemical processes must also be taken into account (see chapter 3). This approach is basic and can be applied to most systems, including chemical reactions in reactors or biological and chemical processes in streams, lakes, and estuaries.

Common Environmental Models

The fundamentals of material transport were introduced in chapter 3, and in this section we show how to apply those fundamentals to the construction of simple models. This section is intended only to be an introduction, and the reader is referred to texts by Thomann [1] and Rich [2] for a more complete treatise on assembling models.

We generated an expression in chapter 3 for the total time derivative or the time rate of change in concentration of material A. This was given in equation (3.36) as

$$\frac{DC_A}{Dt} = \mathscr{D}_{AM}\frac{\partial^2 C_A}{\partial x^2} + R_A$$

Recall that R_A was equal to the molar production of A (moles l^{-1} sec^{-1}) in the volume element. In reality R_A reflects all the sinks and sources of A that will enter in the continuity equation. This is the term that this chapter will evaluate to determine its components for some common systems.

Dissolved Oxygen System

Many of the complex perturbations that man is capable of inflicting on water systems result in a change in the oxygen balance of the system. Thus the dissolved oxygen content of surface water is often used as an index of water quality. Dissolved oxygen models are perhaps the most common models currently in use and therefore will be our first example.

Streeter and Phelps first derived the dissolved oxygen model to predict resultant oxygen concentrations in the Ohio River. The model was used in conjunction with sewage discharges and generated an oxygen concentration versus distance profile known as the "oxygen sag curve." Their basic model has since been refined as more parameters affecting the oxygen in a system have been defined.

The present-day version of this model is written as a mass balance on the oxygen in a system. As was described earlier, the mass balance model includes inputs, outputs, and reaction rates. The input mechanisms (sources) are discussed in the sections that follow.

Reaeration. Water undersaturated with oxygen with respect to the atmosphere will undergo reaeration according to the equation

$$R_1 = k_2(C_S - C) \tag{4.1}$$

where C = concentration of dissolved oxygen in solution at time t (mgl^{-1})

C_S = saturation concentration of oxygen in equilibrium with atmosphere (mgl^{-1})

k_2 = reareation coefficient (t^{-1})

Photosynthesis. Plant material growing in the region where the model is being applied may contribute large amounts of oxygen to the system during daylight hours. Many variables such as nutrients, light, temperature, etc. affect the rate of photosynthesis, hence the rate of oxygen production. Several functions have been employed in the dissolved oxygen model depending on which parameter is limiting the rate of oxygen production in a particular system. For example, when sewage has been introduced to a river or estuary, nutrients such as nitrogen and phosphorus are seldom limiting the rate of oxygen production. Therefore, light is often considered the rate limiting parameter and functions such as

$$R_2 = \begin{cases} \left(\frac{DC}{DT}\right)_{max} \sin \frac{2\Pi(t - t_r)}{(t_s - t_r)} & (t_r < t < t_s) \\ 0 & (t_r > t > t_s) \end{cases} \quad (4.2)$$

are often used where

$\frac{DC}{Dt}$	=	rate of photosynthetic oxygen production at time t ($mgl^{-1}t^{-1}$)
$\left(\frac{DC}{Dt}\right)_{max}$	=	maximum rate of photosynthetic oxygen production ($mgl^{-1}t^{-1}$)
t_r	=	time of sunrise
t_s	=	time of sunset

Reaeration and photosynthesis contribute the largest input of oxygen to streams or estuaries and are usually the only input variables included in the model. The oxygen sinks (oxygen removal) must now be added to complete our mass balance.

Biochemical Oxygen Demand. Perhaps the most significant oxygen removal mechanism in a polluted stream is the oxygen required by microorganisms to stabilize organic matter. This demand on the oxygen can be exerted by suspended material throughout the water column or by sedimented material which exerts a demand on the overlying water.

The rate of oxygen utilization for stabilization of organic matter by microorganisms follows a first-order relationship—that is, the rate is directly proportional to the amount of biodegradable material left in the system.

$$R_3 = -k_1 L \quad (4.3)$$

where k_1 = deoxygenation constant (t^{-1})

L = first-stage BOD remaining after time t (mgl^{-1})

Plant Respiration. Plant material respires (uses oxygen) as well as photosynthesizes. The rate of oxygen respiration appears to be relatively constant throughout a twenty-four hour period, hence a constant rate term may be used in the model. However, because both photosynthesis and respiration are difficult to quantify in any particular system, a combined rate term reflecting both activities is often employed. Rich defines this term as S_R, and the rate equation is given by

$$R_4 = -S_R \tag{4.4}$$

where S_R = rate at which dissolved oxygen changes as a result of photosynthesis and respiration ($mgl^{-1}t^{-1}$).

We have now defined the principle sinks and sources in our mass balance on oxygen. If we now consider a reach of a stream and sum the inputs and outputs of oxygen for that reach, a predictive model can be generated. The time rate of change of oxygen in an element of fluid moving through our reach can be defined as follows:

$$\frac{DC}{Dt} = \mathscr{D}_{CM}\frac{\partial^2 C}{\partial x^2} + R\ (\text{total}) \tag{4.5}$$

or

$$\frac{DC}{Dt} = \mathscr{D}_{CM}\frac{\partial^2 C}{\partial x^2} + k_2(C_S - C) - k_1 L - S_R \tag{4.6}$$

The reader here will recall that we are using the total time derivative following the motion. It is convenient to express the total derivative in equation (4.5) in terms of the partial derivatives of space and time [see equation (3.6)]

$$\frac{\partial C}{\partial x}V_x + \frac{\partial C}{\partial t} = \mathscr{D}_{CM}\frac{\partial^2 C}{\partial x^2} + k_2(C_S - C) - k_1 L - S_R \tag{4.7}$$

Here we are only concerned with the x component of velocity.

We further assume the volumetric flow rate and cross-sectional area of our reach remains constant, so that $V_x = Q/A$ and the section is completely mixed—that is $\mathscr{D}_{CM}\,\partial^2 C/\partial x^2 = 0$—therefore,

$$\frac{\partial C}{\partial t} = \frac{-Q}{A} \times \frac{\partial C}{\partial x} + k_2(C_S - C) - k_1 L - S_R \tag{4.8}$$

where Q = volumetric flow rate ($cm^3\ t^{-1}$)

A = cross-sectional area of stream (cm^2)

The term S_R may have an associated positive or negative sign depending on whether the rate of oxygen removal is larger or smaller than the rate of addition. Equation (4.8) exhibits the negative sign; that is, the rate of addition is greater than the rate of removal.

Due to the dependence of oxygen solubility on salinity and temperature, equation (4.8) is often written in terms of the oxygen deficit. The deficit is defined as the difference between the oxygen saturation value and the concentration at any time t; that is,

$$D = C_S - C \tag{4.9}$$

Thus equation (3.8) becomes

$$\frac{\partial D}{\partial t} = \frac{-Q}{A} \times \frac{\partial D}{\partial x} + -k_2 D + k_1 L + S_R \tag{4.10}$$

At steady-state conditions, the rate of change of oxygen with time is equal to 0; that is, $\partial C/\partial t = 0$ and $\partial D/\partial t = 0$. Equations (4.8) and (4.10) then reduce to

$$\frac{Q}{A} \times \frac{dC}{dx} = k_2(C_S - C) - k_1 L - S_R \tag{4.11}$$

and

$$\frac{Q}{A} \times \frac{dD}{dx} = -k_2 D + k_1 L + S_R \tag{4.12}$$

Equations (4.11) and (4.12) reflect the change in oxygen concentration with distance down a stream reach that might be expected due to the included parameters. These equations must be integrated to yield a relationship for oxygen deficit or concentration versus distance. The mathematical approach for treating these differential equations is given in chapter 5.

Before leaving our dissolved oxygen model, it is useful to note that often a mass balance is written on the organic material throughout the reach, which regulates the oxygen concentration. This is done to afford engineers a model that will predict the allowable organic loading (BOD) to maintain a required dissolved oxygen content in the stream. This is done by applying equation (4.12) to the movement of organic matter (L) instead of dissolved oxygen. In this case we must also introduce two new rate terms which affect the movement of BOD in the stream section. The two new rate terms are:

L_a = rate of addition of BOD to the stream by nonpoint sources. ($mgl^{-1}\ t^{-1}$)

$k_3 L$ = rate of BOD removal due to sedimentation ($mgl^{-1}\ t^{-1}$)

It can be shown that

$$\frac{\partial L}{\partial t} + \frac{Q}{A} \times \frac{\partial L}{\partial x} = -k_1 L - k_3 L + L_a \tag{4.13}$$

At steady state,

$$\frac{Q}{A} \times \frac{\partial L}{\partial x} = -(k_1 + k_3)L + L_a \tag{4.14}$$

Equation (4.14) can then be integrated, and a spatial distribution of BOD over the reach can be determined. These values can be substituted into equation (4.12), and the typical dissolved oxygen sag curve will be generated (see chapter 5).

Eutrophication Models

Models that attempt to simulate the eutrophication process have recently attracted much attention. These particular models vary with respect to the particular parameter studied. Many contain mass balance relationships written on the macronutrients, while others model the production of phytoplankton biomass with time. As was mentioned earlier, most all of these models utilize the same basic concept—that is, conservation of mass.

Many of the phytoplankton production models that are now appearing in the literature are reworked ecological succession models. Chen [3] has reviewed many of these models, and the reader is referred to his paper for general background information. Perhaps one good example of an algal productivity model is that proposed by Lorenzen and Mitchell [4]. They derived a relationship to predict the resultant algal biomass after destratification of deep reservoirs. Once again it is seen that the same basic approach is used to develop the model, and in chapter 5 we show how to treat the model mathematically.

This particular model assumes light is limiting the rate of algal growth, and the rate of change of algal concentration is given by

$$\frac{DC}{Dt} = R_C \tag{4.15}$$

or

$$\frac{DC}{Dt} = kC - rC \tag{4.16}$$

where C = algal concentration (mgl^{-1})

k = specific growth rate as a function of light (t^{-1})

r = specific rate of respiration (t^{-1})

t = time

Here again we assume diffusion—that is,

$$\mathscr{D}_{CM}\frac{\partial C^2}{\partial x^2} = 0$$

Specific functions can now be substituted into the basic mass balance equation [equation (4.16)]. Because the above model is dependent on light intensity, a rate constant such as that described by Smith [5] may be used

$$k = \frac{k_{\max} A\ I}{[I + (A\ I)^2]^{1/2}} \tag{4.17}$$

where k = specific growth rate (t^{-1})

$k_{\max}$ = specific growth rate at light saturation (t^{-1})

A = constant (Lux^{-1})

I = light intensity (Lux)

Thus the model now reflects the rate of change of algal biomass as a function of light intensity. However, because the model is to predict resultant algal biomass concentrations as a function of mixing depth, some function relating specific growth rate to depth is required. Lorenzen approximates this function with the equation

$$I_d = I_0 e^{-(\alpha+\beta C)d} \tag{4.18}$$

where I_d = light intensity at any depth, d (Lux)

I_0 = light intensity at the water surface (Lux)

α = light attenuation coefficient of the medium (m^{-1})

β = light attenuation coefficient due to algal cells ($\text{m}^{-1}\ \text{mg}^{-1}\ \text{l}$)

C = concentration of algal cells (mgl^{-1})

d = depth (m)

Equations (4.17) and (4.18) can now be substituted into equation (4.16), yielding the following relationship for the instantaneous rate of algal production at any depth

$$\frac{DC}{Dt} = \frac{k_{\max} A\ I_0 e^{-(\alpha+\beta C)d}}{[1 + (A\ I_0 e^{-(\alpha+\beta C)d})^2]^{1/2}} C - rC \tag{4.19}$$

This total time derivative may be expressed in terms of the partial derivatives of time and space as

$$\frac{\partial C}{\partial t} + V_x\frac{\partial C}{\partial x} + V_y\frac{\partial C}{\partial y} + V_z\frac{\partial C}{\partial y} = \frac{k_{\max} A\ I_0 e^{-(\alpha+\beta C)d}}{[1 + (A\ I_0 e^{-(\alpha+\beta C)d})^2]^{1/2}} C - rC \tag{4.20}$$

In this case it is assumed there is no water movement in the lake, hence the velocity terms are 0. Thus the total time derivative in equation (4.19) may simply be replaced with the partial time derivative to yield

$$\frac{\partial C}{\partial t} = \frac{k_{max} A\, I_0 e^{-(\alpha+\beta C)d}}{[1 + (A\, I_0 e^{-(\alpha+\beta C)d})^2]^{1/2}}\, C - rC \qquad (4.21)$$

Equation (4.21) relates the instantaneous rate of algal production at any depth d. A useful form of equation (4.21) is obtained if it is integrated over the depth of the water column ($d = 0$ to $d = z$). It can then be used for the determination of the instantaneous algal production throughout the entire water column.

Biological Processes in Reactors

This section is concerned with the behavior of certain biological processes in reactors. It is intended to show how environmental processes may proceed in reactors and be defined mathematically. For a treatise on reactor kinetics, the reader is referred to Smith [6].

Perhaps the most simple model is that of a continually stirred flow through tank reactor that has a biological process actively progressing within it. Again the general model starts with a mass balance of the cellular material in the reactor:

$$V \frac{dC}{dt} = QC_0 - QC + R_g V - R_d V \qquad (4.22)$$

where Q = flow (ml min^{-1})

C_0 = concentration of cellular material in the influent (cells ml^{-1})

C = concentration of cellular material in the reactor (cells ml^{-1})

R_g = volumetric growth rate of microorganisms (cells ml^{-1} min^{-1})

R_d = volumetric decay rate of microorganisms (cells ml^{-1} min^{-1})

V = volume of reactor (ml)

Quite often the above process is run at steady state; that is, $dC/dt = 0$. When this occurs, equation (4.22) reduces to the following linear algebraic equation, which may be solved directly:

$$QC_0 - QC + R_g V - R_d V = 0 \qquad (4.23)$$

In the case where the influent flow does not contain cellular material—that is, $C_0 = 0$—and the growth rate terms (R_g and R_d) are replaced with the specific growth constants,

$$R_g = \mu C \tag{4.24}$$

$$R_d = \sigma_d C \tag{4.25}$$

where μ = specific growth constant (t^{-1})

σ_d = specific death constant (t^{-1})

equation (4.23) becomes

$$-QC + \mu CV - \sigma_d CV = 0 \tag{4.26}$$

or

$$\mu - \sigma_d = \frac{Q}{V} = \frac{1}{\tau} \tag{4.27}$$

where τ = hydraulic residence time (t)

Equation (4.27) represents the basic chemostat or algastat operating at steady state. These reactors have proven very useful for evaluating microbial growth kinetics as a function of nutrient enrichment. The reader should note the simplicity of the solution when the reactor was running at steady state. Waite, et al. [7] investigated the mathematics resulting from a chemostat not running at steady state—that is, $dC/dt \neq 0$. This resulted from the fact that no material was allowed to wash out of the reactor. Because the biomaterial in the reactor was not unicellular, a mass balance was written on the oxygen production which was directly proportional to algal growth, yielding

$$V\frac{dC}{dt} = QC_0 - QC + R_gM - R_dM \tag{4.28}$$

where C = concentration of dissolved oxygen in reactor (mgl^{-1})

M = algal mass (mg)

C_0 = concentration of dissolved oxygen in the influent stream (mgl^{-1})

and Q, R_g, R_d, t, and V are as defined in the previous example.

As we saw in chapter 2, algal growth can be assumed to be first order: that is,

$$\frac{dM}{dt} = (k_LR_g - k_dR_d)M \tag{4.29}$$

where k_L = stoichiometric growth constant [mg(mg O_2)$^{-1}$]

k_d = stoichiometric decay constant [mg(mg O_2)$^{-1}$]

This may be solved to yield (see chapter 5)

$$M = M_0 e^{(k_L R_g - k_d R_d)t} \tag{4.30}$$

Equation (4.30) may now be substituted into equation (4.29) to yield

$$V\frac{dC}{dt} = QC_0 - QC + (R_g - R_d)M_0 e^{(k_L R_g - k_d R_d)t} \tag{4.31}$$

Equation (4.31) now relates the increase in biomass in a reactor with no biomaterial washout. The solution to this differential equation is shown in chapter 5.

Other Models

Many other environmental systems have been modeled, and each exhibits its own specific combination of rate terms. Another system worth reviewing is that of microbial die-off in natural systems.

Coliform bacteria are used as indicators of pollution by warm-blooded animals. The fate of these indicator organisms in the environment is therefore very important. Several environmental factors contribute to bacterial decline, including solar radiation, sedimentation, predation, toxins, etc. Many models have been conceived to predict bacterial concentrations with time (die-off), and the reader is referred to the work of Mitchell and Chamberlin [8] for an overview of the field.

As with the case of our dissolved oxygen model, a microbial die-off model may be constructed by adding together the various factors affecting microbial viability in a control volume. The most simple and commonly used relationship for microbial die-off due to physiochemical processes is Chick's law:

$$\frac{DC}{Dt} = -k_c C \tag{4.32}$$

where C = concentration of bacterial cells (mgl^{-1})

k_C = specific die-off constant (t^{-1})

This is a simple first-order relationship, showing that the rate of change of bacteriological concentration is dependent only on the concentration of bacteria present.

More specific relationships have been derived for the individual die-off mechanisms. Pirt [9] has suggested that the mathematical representation of microbial die-off due to nutrient deficiencies takes the form:

$$\frac{DC}{Dt} = \left(\frac{\mu_{max} S}{K_s + S} - a\right)C \tag{4.33}$$

where μ_{max} = maximum growth rate (t^{-1})

S = substrate concentration (mgl^{-1})

K_s = half saturation constant (mgl^{-1})

a = specific maintenance constant (t^{-1})

C = concentration of bacteria (mgl^{-1})

Note that Pirt has incorporated Monod enzyme kinetics into the model. The inclusion of rational functions such as the Monod equation in models has been treated in chapter 2.

Another aspect of microbial die-off is the solar radiation component. Gameson and Gould [10] have postulated that die-off of enteric bacteria due to sunlight can be represented by the exponential function in the following manner:

$$\frac{DC}{Dt} = -k_L I_0 e^{-\alpha z} C \tag{4.34}$$

where C = concentration of bacteria (mgl^{-1})

k_L = proportionality coefficient (Lux^{-1})

I_0 = light intensity below surface of water (Lux)

α = attenuation coefficient (M^{-1})

z = depth (m)

This function would have to be integrated over depth z to find the remaining bacterial concentration due to sunlight effects.

As with the dissolved oxygen model, the above components can be combined, and a predictive model for microbial die-off due to environmental factors can be created:

$$\frac{DC}{Dt} = -k_c C + \left(\frac{\mu_{max} S}{K_s + S} - a\right) C - k_L I_0 e^{-\alpha z} C \tag{4.35}$$

Once again, in order to treat equation (4.35), the total time derivative must be reduced to partial differentials for simplification. This yields

$$\frac{\partial C}{\partial t} + V_x \frac{\partial C}{\partial x} + V_y \frac{\partial C}{\partial y} + V_z \frac{\partial C}{\partial z} = -k_c C + \left(\frac{\mu_{max} S}{K_s + S} - a\right) C - k_L I_0 e^{-\alpha z} C \tag{4.36}$$

Equation (4.36) can now be simplified according to certain physical constraints in the system being modeled. Examples of how the above model may be treated are given in chapter 5.

In this chapter we have seen how many different environmental models are generated from the same basic principles. The concept of conservation of mass applied to the various elements of environmental systems has been the underlying framework of all our models, and in like manner we have generated the same type of differential equation in each case. We describe how these common differential equations are treated in chapter 5.

References

1. Thomann, R. V. (1972), *Systems Analysis and Water Quality Management*, New York, McGraw-Hill.
2. Rich, L. G., (1973), *Environmental Systems Engineering*, New York, McGraw-Hill.
3. Chen, C. W. (1970), *J. San. Eng. Div.* (Am. Soc. Civil Eng.) 96: 1085.
4. Lorenzen, M., and R. Mitchell (1973), *Environmental Science and Technology* 7, no. 10.
5. Smith, E. L. (1936), *Proc. Nat. Acad. Sci.* (Washington, D.C.), 22: pg. 504.
6. Smith, J. M. (1970), *Chemical Engineering Kinetics*, New York, McGraw-Hill.
7. Waite, T., L. Spielman, and R. Mitchell (1972), *Environmental Science and Technology* 6, no. 13.
8. Mitchell, R., and C. Chamberlin (1974), *Proc. Int. Symp. on Discharge of Sewage from Sea Outfalls* (London): 25.
9. Pirt, S. J. (1965), *Proc. R. Soc. B.* 163.
10. Gameson, A. L., and D. J. Gould (1974), *Proc. Int. Symp. on Discharge of Sewage from Sea Outfalls* (London): 22.

5 Differential Equations

Introduction

We are concerned in this chapter with solution techniques to differential equations common to environmental processes. A *differential equation* is defined as an equation which involves derivatives of an unknown function. Thus the equation

$$\frac{dC}{dx} = x \tag{5.1}$$

is a differential equation. From this equation the unknown function C can be determined. As an example the reader should have no difficulty in selecting a function C whose first derivative is equal to x. For example, the function $C = x^2/2$, when substituted into the left-hand side of this equation, yields an identity. We, therefore, call this function a *solution*, and we say that it *satisfies* the differential equation. We also note here that the function $C = x^2/2 + K$, where K is any constant, is also a solution (since $dK/dx = 0$). It will be seen that to determine the unique solution to a differential equation certain conditions of that differential equation must be specified. The process used in solving this differential equation is called *antidifferentiation*. It is the reverse of differentiation, and slightly more difficult.

A generalization of equation (5.1) is the equation

$$\frac{dC}{dx} = f(x) \tag{5.2}$$

Whereas it was easy to determine the antiderivative for the special case $f(x) = x$, this is in general not true for most forms of $f(x)$. Recall the familiar symbol

$$\int f(x)\, dx$$

which signifies the *indefinite integral* of $f(x)$. The indefinite integral is the most general antiderivative of $f(x)$. For $f(x) = x$, the indefinite integral is

$$\int x\, dx = \frac{x^2}{2} + K \tag{5.3}$$

where K is an arbitrary constant. We express the solution to the equation

$$\frac{dC}{dx} = f(x)$$

as

$$C = \int f(x)\, dx + K$$

Another type of differential equation,

$$\frac{dC}{dx} = kC \tag{5.4}$$

was introduced in chapter 2. The solution to this equation can be shown to be

$$C = Ke^{kx} \tag{5.5}$$

where K is an arbitrary constant. We stated in chapter 2 that the only *function* which satisfies the relationship $dC/dx = kC$ is the special exponential function $C = e^{kx}$. The reader should verify that multiplying e^{kx} by K produces a function that still satisfies the differential equation. The multiplicative constant K converts the function e^{kx} from *an* antiderivative to the *most general* antiderivative.

Several other types of differential equations were introduced in chapter 3 while discussing material transport, and in chapter 4 when mathematically modeling environmental processes. In the following sections we introduce the reader to means of identifying, classifying, and solving those differential equations discussed previously, and to others that may be of interest. These examples illustrate the need for a knowledge of the indefinite integrals of certain functions for the solution to a particular type of differential equation. This knowledge of integration (finding the indefinite integral) will also be invaluable for the solution of differential equations whose solutions are not directly integrable. For this reason the first section of this chapter concerns the indefinite integrals of some of the more common functions. The relationship between integration and the solution to differential equations certainly suggests why the term *integral* is often used in lieu of *solution*.

Indefinite Integrals

We have defined the *indefinite integral* of $f(x)$

$$\int f(x)\, dx$$

as the most general *antiderivative* of $f(x)$. In calculus integration is usually related to concepts such as area and volume. The reader may want to pursue this path to refresh his memory on this important subject. We, however, wish to discuss *integration* as a process of determining a function whose derivative is given (antidifferentiation). Accordingly, we give in table 5–1 a list of common formulas and rules of integration. The reader should note the inverse processes of differentiation and indefinite integration by comparing tables 5–1 and 2–1.

Identification and Classification of Equations

We have defined differential equations as equations in which derivatives of an unknown function appear. Definitions similar to this and others more precise (and complex) appear throughout the literature. While a more mathematically articulate definition could be given here, we felt that in keeping with the general tenor of this book, certain examples might better convey the meaning.

Consider the differential equations

$$\frac{dC}{dx} = f(x) \tag{5.6}$$

and

$$\frac{dC}{dx} = kC \tag{5.7}$$

In accordance with our definition, note the derivatives of an unknown function C appearing in the equations. This dependent variable C is a function of the independent variable x. Typically, a differential equation is valid for specified values of the independent variable—that is, the equation would be defined over an interval; for example $a < x < b$.

Type

Those differential equations that contain derivatives with respect to only one dependent variable are called *ordinary* differential equations. In addition to equations (5.6) and (5.7) the following equations are ordinary

$$\frac{d^2C}{dx^2} = f(C,x) \tag{5.8}$$

$$\frac{dC}{dx} + 2C = e^x \tag{5.9}$$

Table 5–1
Formulas and Rules of Integration

Function	*Indefinite Integral*[a]	
$C = k$	$\int C\,dx = kx$	(k = constant)
$C = x^a$	$\int C\,dx = (x^{a+1})/(a+1)$	($a \neq -1; x \neq 0$ if $a < 0$)
$C = k\,f(x)$	$\int C\,dx = k \int f(x)\,dx$	
$C = f(x) + g(x)$	$\int C\,dx = \int f(x)\,dx + \int g(x)\,dx$	(g = function)
$C = f(x)\,g'(x)$	$\int C\,dx = f(x)\,g(x) - \int g(x)\,f'(x)\,dx$	
$C = dx/x$	$\int C\,dx = \ln x$	
$C = [f'(x)]/[f(x)]$	$\int C\,dx = \int (df/f) = \ln f(x)$	
$C = f^a(x)\,f'(x)$	$\int C\,dx = [f(x)^{a+1}]/(a+1)$	
$C = e^x$	$\int C\,dx = e^x$	
$C = e^{f(x)}\,f'(x)$	$\int C\,dx = e^{f(x)}$	

[a] It is understood that to each integral an arbitrary constant of integration (K) is to be added.

$$\left(\frac{dC}{dx}\right)^2 + 2 \ln C = x \tag{5.10}$$

$$C\left(\frac{d^3C}{dt^3}\right)^{1/2} + 3\left(\frac{d^2C}{dt^2}\right)^2 = 0 \tag{5.11}$$

The salient features of each of the above equations are (a) the derivative of an unknown function C, and (b) only one independent variable appears. We point out here that the existence in these equations of higher order derivatives of C, powers of C or its derivatives, and the appearance of other known functions are all permitted within our definition of ordinary differential equations.

If there are two or more independent variables, partial derivatives must appear, and we call the equation a *partial* differential equation. Thus

$$\frac{\partial C}{\partial t} + 2\frac{\partial C}{\partial x} = te^{x+3t} + \frac{\partial^2 C}{\partial y^2} \tag{5.12}$$

$$\frac{\partial^2 C}{\partial t^2} + \frac{\partial^2 C}{\partial x^2} = 0 \tag{5.13}$$

$$C\left(\frac{\partial C}{\partial t}\right) + \left(\frac{\partial C}{\partial x}\right)^2 = xt \tag{5.14}$$

are partial differential equations. In equation (5.12) the dependent variable C is a function of three independent variables x, y, and t. In equations (5.13) and (5.14) C is a function of the independent variables x and t.

Sometimes a physical problem generates *simultaneous* differential equations in which two or more *dependent* variables appear. For example,

$$\frac{\partial C_A}{\partial t} + 2\frac{\partial C_B}{\partial t} + C_B = t^{-1}$$

and

$$\frac{\partial C_A}{\partial t} + C_A + 3\ C_B = e^{-t}$$

involve the dependent variables C_A and C_B and the single independent variable t.

Order and Degree

The *order* of a differential equation is given by the highest order derivative appearing in the equation. Thus the equations

$$\frac{dC}{dx} = f(c,x) \tag{5.15}$$

$$\left(\frac{dC}{dx}\right)^3 + \frac{d^2C}{dx^2} + C^2 = \ln x \tag{5.16}$$

$$\frac{\partial^4 C}{\partial x^4} + 2\ \frac{\partial^4 C}{\partial x^2 \partial t^2} + C\ \frac{\partial^4 C}{\partial t^4} = \left(\frac{\partial C}{\partial y}\right)^5 + Ce^t \tag{5.17}$$

are of order one, two, and four, respectively.

The *degree* of a differential equation is the exponent of the highest order derivative that appears. This definition assumes that the equation has been cleared of radicals and fractions with respect to all derivatives of the dependent variable. Thus equations (5.15), (5.16), and (5.17) are all of degree one. The equation

$$\left(\frac{d^2C}{dx^2}\right)^2 + \left(\frac{dC}{dx}\right)^3 = C^5 + \ln x \tag{5.18}$$

is of degree two, while the equation

$$\sqrt{\left(\frac{dC}{dt}\right)^5} = \frac{d^2C}{dt^2} + C \tag{5.19}$$

can be cleared of the radical by squaring both sides to yield the second-degree equation

$$\left(\frac{dC}{dt}\right)^5 = \left(\frac{d^2C}{dt^2}\right)^2 + 2C\left(\frac{d^2C}{dt^2}\right) + C^2$$

The equations

$$\frac{d^2C}{dt^2} + \sin C = 0 \tag{5.20}$$

and

$$\frac{dC}{dx} + \ln C = 0 \tag{5.21}$$

are examples of differential equations for which a degree is not definable. This is due to the presence of a nonalgebraic function.

Linear Differential Equations

Differential equations in a dependent variable C are *linear* if the *exponent* of C and *all derivatives* of C present in the equation are *equal to one*—that is, they are linear if the equation is of degree one for C and its derivatives. Linear differential equations are often used to model physical systems because of their relative simplicity of solution as compared with nonlinear differential equations. The following equations taken from previous examples are linear (degree = 1)

$$\frac{dC}{dx} = f(x) \tag{5.22}$$

$$\frac{dC}{dx} = kC \tag{5.23}$$

$$\frac{dC}{dx} + 2C = e^x \tag{5.24}$$

$$\frac{\partial C}{\partial t} + 2\frac{\partial C}{\partial x} = te^{x+3t} + \frac{\partial^2 C}{\partial y^2} \tag{5.25}$$

$$\frac{\partial^2 C}{\partial t^2} + \frac{\partial^2 C}{\partial x^2} = 0 \tag{5.26}$$

Examples of nonlinear differential equations, also discussed previously, are

$$\left(\frac{dC}{dx}\right)^2 + 2\ln C = x \tag{5.27}$$

$$C\left(\frac{d^3C}{dt^3}\right)^{1/2} + 3\left(\frac{d^2C}{dt^2}\right)^2 = 0 \tag{5.28}$$

$$C\left(\frac{\partial C}{\partial t}\right) + \left(\frac{\partial C}{\partial x}\right)^2 = xt \tag{5.29}$$

$$\left(\frac{dC}{dx}\right)^3 + \frac{d^2C}{dx^2} + C = \ln x \tag{5.30}$$

$$\frac{dC}{dx} + \frac{d^3C}{dx^3} + C^2 = k \tag{5.31}$$

Note that terms like C^2, $(dC/dx)^2$, $C(\partial C/\partial t)$, and $\ln C$ are all nonlinear.

Homogeneous Linear Differential Equations

A *linear* differential equation in a dependent variable C is *homogeneous* if each and every term of the equation is of the first degree in C and its derivatives, excluding the zero term. A general form of a linear first-order differential equation is

$$\frac{dC}{dx} + P(x)C = Q(x) \tag{5.32}$$

This equation is called homogeneous if $Q(x) = 0$. Similarly, the linear second-order differential equation

$$\frac{d^2C}{dx^2} + P_1(x)\frac{dC}{dx} + P_2(x)C = Q(x) \tag{5.33}$$

is also homogeneous if $Q(x) = 0$. The reader should be able to extend this line of reasoning for equations of higher order.

Specific examples of homogeneous differential equations are given by

$$\frac{dC}{dx} - kC = 0 \tag{5.34}$$

and

$$\frac{d^3C}{dx^3} + e^x\frac{d^2C}{dx^2} + x^2 \ln x \frac{dC}{dx} + \sin x(C) = 0 \tag{5.35}$$

and of nonhomogeneous differential equations by

$$\frac{dC}{dx} - kC = e^x \tag{5.36}$$

and

$$\frac{d^2C}{dx^2} = x^3 \tag{5.37}$$

There appears to be a general acceptance throughout many standard textbooks that the term *homogeneous* be used to define two different concepts. Later in this chapter when we discuss solution techniques, another definition of the term homogenous appropriate to first-order differential equations is given.

Solution Techniques

The first discussion of the concept of solution techniques for differential equations was given in the introduction to this chapter, where it was related to the special equation $dC/dx = f(x)$. We define a *solution* to a differential equation in which C is the unknown dependent variable as any relationship for C [such as $C = f(x)$ or $F(C,x) = 0$] which, upon substitution into the equation, yields an identity. The function $C(x)$ is then said to satisfy the differential equation. It is easy to see that the function $C = f(x) = e^{kt}$ satisfies the equation

$$\frac{dC}{dx} = kC \tag{5.38}$$

Multiplication by an arbitrary constant K yields an infinitude of solutions Ke^{kx}, a characteristic of differential equations.

The reader should have formulated a host of questions by this time about differential equations and their solution, such as:

1. Are the dependent variables, like the independent variables of a differential equation, constrained to a particular interval of numbers?
2. How do we know that a solution to the differential equation exists?
3. If there is more than one solution, how is the solution appropriate to the physical problem at hand selected?
4. What types of differential equations are common to a specific physical system?
5. What are the methods most often useful in solving specific types of differential equations?

Complete answers to some of the foregoing questions are beyond the scope or intention of this book, but for the purposes of guidance, the following discussion to each is given:

1. We often discuss a differential equation and its solution without giving

a range for the independent variable(s) or a domain for the dependent variable. This approach lends itself to a certain conciseness and flexibility in presentation. In a given problem these particular intervals are often obvious and are usually dictated by the geometry or some other notion related to the physical system.

2. This important question refers to the concept of *existence* in differential equations, and is beyond the scope of this text. We assume that solutions to the equations we treat do exist and we proceed accordingly. This precludes our examining equations such as $(dC/dx)^2 = -1$, which has no *real* solution, and $|dC/dx| = -1$, which has no solutions, real or nonreal.
3. The concept of *uniqueness* is suggested by this question and is of utmost importance in our treatment of differential equations. Suppose the solution to a specific differential equation, which modeled a physical system, was not unique. This would allow for the possibility of two or more investigators arriving at different solutions to the same equation and thus predicting different behavior for the same physical system. In our treatment of differential equations a specific, unique solution is isolated by constraining the solution to satisfy additional conditions associated with the physical problem. These conditions are called *boundary* or *initial conditions* and are discussed in more detail in subsequent sections.
4. Those differential equations of most interest to us are suggested by the title to this section and were discussed under the section on linearity. Linear equations are those that are most often encountered in environmental modeling and must certainly be well understood before undertaking a treatment of nonlinear differential equations.
5. There are numerous methods, which we call *solution techniques*, and many are classical. This chapter is devoted principally to the discussion of certain of these solution techniques. The reader should know that there may be more than one solution technique available to solve a single differential equation.

First-Order Equations

We first apply various solution techniques to differential equations of the first order. Recall that a first-order equation contains only first-order derivatives of the unknown dependent variable. The most general first-order differential equation has the form

$$\frac{dC}{dx} = f(C,x) \tag{5.39}$$

of which the previously defined equations

$$\frac{dC}{dx} = f(x) \tag{5.40}$$

$$\frac{dC}{dx} = kC \tag{5.41}$$

$$\frac{dC}{dx} + 2C = e^x \tag{5.42}$$

and

$$\frac{dC}{dx} + \ln C = 0 \tag{5.43}$$

are special cases.

Another standard form of the general first-order differential equation, which gives equal importance to the two variables present, is

$$M(x,C)\,dx + N(x,C)\,dC = 0 \tag{5.44}$$

where dx and dC are called the *differentials* of x and C, respectively. The essential equivalency of these two forms can be illustrated by algebraic manipulation of an equation represented by equation (5.39) to one represented by equation (5.44). We use the term *essential equivalency* because equation (5.39) does not always admit all the solutions allowed by equation (5.44). For example, consider the equation

$$\frac{dC}{dx} = +\frac{C}{x} \tag{5.45}$$

This can be written as

$$C\,dx - x\,dC = 0 \tag{5.46}$$

where $M(x,C) = C$ and $N(x,C) = -x$. The reader should be able to see that $x = 0$ is a solution of equation (5.46), but cannot be for equation (5.45) because dC/dx would be undefined at this point. It is of interest to note that equation (5.45) by simple algebra could be put into another form $dC/C = dx/x$, which would exclude the solution $C = 0$, admissable by both equations (5.45) and (5.46). The following example also illustrates that certain solutions of equation (5.44) may not be solutions of equation (5.39). The equation

$$\frac{dC}{dx} = \frac{\cos x}{\sin x}\,C \tag{5.47}$$

can be written as

$$C \cos x\, dx - \sin x\, dC = 0 \tag{5.48}$$

where $M(x,C) = C \cos x$

$N(x,C) = -\sin x$

The special solutions $x = 0, \pm\pi, \pm 2\pi, \ldots$ satisfy equation (5.48) regardless of the value of C but are not admissable for equation (5.47); dC/dx would again be undefined due to the zeros that would appear in the denominator.

These examples reveal the notion of special or extra solutions, often obtained by observation and without integration. These solutions may be overlooked due to the form of the differential equation, but may be critical to an investigator analyzing a physical system. These oversights are often the result of subtle algebraic manipulations in formulating the equation, and we caution the investigator to take great care in his/her analysis to avoid them.

The identification and classification of differential equations should precede attempts at solution as did our discussion of these topics. This will frequently allow for a routine solution of the equation by one of the many standard methods available to the investigator. We suggest s/he become familiar with as many as possible, as there is no best method for the great variety of differential equations that an investigator, even in a specialized field, may encounter.

It is suggested that *guessing* be a part of every investigator's "bag of tricks" in seeking solutions to differential equations. Any technique used in arriving at a solution is valid; however, the solution should always be verified by substitution back into the equation.

Variables Separable. Differential equations can sometimes be separated into terms, each involving only one variable. Thus for the general first-order linear differential equation in the form $dC/dx = f(x,C)$ to be subject to this solution technique, the function $f(x,C)$ would have to have the property that $f(x,C) = g(x)\, h(C)$—that is, $f(x,C)$ can be written as a multiple of a function of x alone by a function of C alone. Thus the equation

$$\frac{dC}{dx} = g(x)\, h(C) \tag{5.49}$$

can be written with *variables separated* as

$$\frac{dC}{h(C)} = g(x)\, dx \tag{5.50}$$

(This is exactly the type of algebraic manipulation that could disguise solutions to the original equation as discussed in the previous section.) Each term of this equation is a function of only one variable, and thus the solution has been reduced to the determination of their indefinite integrals. We write this as

$$\int \frac{dC}{h(C)} = \int g(x)\,dx + K \tag{5.51}$$

where K is an arbitrary constant of integration.

The solution is by no means completed at this point, since the evaluation of the integrals may not be in terms of elementary functions, and special functions, or numerical or other types of integration, may be required.

Example 5.1. A first-order reaction is one in which the concentration of only one reacting substance changes. The rate of change is proportional to the amount of that material present at any instant. Expressed mathematically, we have the equation

$$\frac{dC_A}{dt} = kC_A$$

where C_A = concentration of material A

k = proportionality constant

Although we have encountered this equation and know its solution, we will proceed in ignorance and separate the variables, yielding

$$\frac{dC_A}{C_A} = k\,dt$$

Note here that the original equation admitted the solution $C_A = 0$, while in the variables separated form it does not. This solution can be readily discounted, however, by physical reasoning. Integrating each term we get

$$\int \frac{dC_A}{C_A} = \int k\,dt$$

or

$$\ln C_A = kt + K$$

recalling that

$$e^{\ln x} = x$$

we put both sides of the equation in exponential form, or

$$C_A = e^{kt+K} = e^K e^{kt} = \bar{K}e^{kt}$$

where we have written $e^K = \bar{K}$, a more conventional form for the arbitrary constant.

We see that $\bar{K}e^{kt}$ is the solution to the original differential equation. For a specific physical problem, we would have to know the value of the concentration of A at some time to determine the constant $\bar{K}$.

For example, suppose that at $t = 0$ and $C_A = 2$, substituting into the solution $C_A = \bar{K}e^{kt}$ the known value of C_A at $t = 0$ yields

$$2 = \bar{K}e^{k \times o}$$

thus

$$\bar{K} = 2$$

The solution is now completely determined as

$$C_A = 2e^{kt}$$

The other form of the general first-order linear differential equation [equation (5.44)], $M(x,C)\,dx + N(x,C)\,dC = 0$ can be put in variables separated form for the special case $M = M(x)$ and $N = N(C)$. The resulting equation is

$$M(x)\,dx + N(C)\,dC = 0 \tag{5.52}$$

where each term is a function of only one variable. We therefore solve this equation in a similar manner to equation (5.50) to obtain

$$\int M(x)\,dx + \int N(C)\,dC = K \tag{5.53}$$

If we write $M(x) = -g(x)$ and $N(C) = 1/h(C)$, we see that equations (5.51) and (5.53) are equivalent.

Example 5.2. At a given instant of time the concentration gradient of material A in a stream is given by

$$\frac{dC_A}{dx} = -C_0 x^{-2}$$

where C_0 is a known constant. It is also known that at $x = 10$, $C_A = 12$. To determine the concentration at a specific location, we first separate the variables to get

$$dC_A = -C_0 x^{-2}\,dx$$

Integrating both sides yields

$$C_A = C_0 x^{-1} + K$$

Substituting in the known value of C_A at $x = 10$ yields the equation

$$12 = C_0 10^{-1} + K$$

The constant K is determined to be $K = 12 - C_0 10^{-1}$, enabling us to write our solution as

$$C_A = 12 + C_0(x^{-1} - 10^{-1})$$

Example 5.3. A second-order chemical reaction may evolve when the rate of disappearance of a compound is dependent on the concentration of the compound squared. Thus a relationship such as shown here may evolve:

$$\frac{dC}{dt} = -k\left(2 - \frac{C}{5}\right)\left(16 - 4\frac{C}{5}\right)$$

where k is a proportionality constant. We also know that at $t = 0$, $C = 5$. Separating variables, we get

$$\frac{dC}{\left(2 - \frac{C}{5}\right)\left(16 - \frac{+C}{5}\right)} = -k\,dt$$

The fraction on the left-hand side may be written in a form more convenient for integration by the method of partial fractions. We write

$$\frac{1}{\left(2 - \frac{1}{5}C\right)\left(16 - \frac{4}{5}C\right)} = \frac{A}{\left(2 - \frac{1}{5}C\right)} + \frac{B}{\left(16 - \frac{4}{5}C\right)}$$

or by multiplying both sides by $(2 - C/5)(16 - 4C/5)$ to get

$$1 = A\left(16 - \frac{4}{5}C\right) + B\left(2 - \frac{1}{5}C\right)$$

This is an identity and true for all values of C; thus first setting $4/5C = 16$ and then $1/5C = 2$, we get $1 = B(-2)$ and $1 = A(8)$ or $B = -1/2$ and $A = 1/8$. The equation can now be written as

$$\frac{\left(\frac{1}{8}\right)dC}{\left(2 - \frac{1}{5}C\right)} + \frac{\left(-\frac{1}{2}\right)dC}{\left(16 - \frac{4}{5}C\right)} = -k\,dt$$

or simplifying

$$\frac{dC}{(10 - C)} - \frac{dC}{(20 - C)} = -\frac{8}{5}k\,dt$$

Integration yields

$$-\ln(10 - C) + \ln(20 - C) = -\frac{8}{5}kt + K$$

For convenience we replace the arbitrary constant K by another equivalent arbitrary constant $\ln K$ and write the solution as

$$\ln\frac{(20 - C)}{K(10 - C)} = -\frac{8}{5}kt$$

or

$$\frac{20 - C}{K(10 - C)} = \exp\left(-\frac{8}{5}kt\right)$$

Solving for C yields

$$C = 10\frac{\left[\bar{K}\exp\left(-\frac{8}{5}kt\right) - 2\right]}{\left[\bar{K}\exp\left(-\frac{8}{5}kt\right) - 1\right]}$$

when $t = 0$, $C = 5$, and we have

$$5 = 10\left(\frac{\bar{K} - 2}{\bar{K} - 1}\right)$$

Thus $\bar{K} = 3$ (or $K = \ln 3$), and the solution becomes

$$C = 10\,\frac{\left[3\exp\left(-\frac{8}{5}kt\right) - 2\right]}{\left[3\exp\left(-\frac{8}{5}kt\right) - 1\right]}$$

Solutions and Arbitrary Constants. The first differential equation we treated in this chapter

$$\frac{dC}{dx} = x \tag{5.54}$$

had an infinite number of solutions

$$C = \frac{1}{2}x^2 + K \tag{5.55}$$

due to the presence of the constant K. We called K an *arbitrary constant* since whatever value it may be given $C = x^2/2 + K$ is still a solution. Arbitrary constants were seen to be a result of the indefinite integration used to determine the solution. The arbitrary constants were introduced into the solutions (as a result of integration) of those differential equations for which the variables were separable.

The process of solving a differential equation will at some point involve integration so that the derivatives or differentials can be eliminated. An arbitrary constant of integration is introduced for each integration required; that is, for each derivative or differential to be removed. An nth order differential equation will therefore in general possess a solution containing n arbitrary constants. This solution is commonly referred to as a *general solution*. Some authors are annoyed by this terminology, and do not care to use the expression "the general solution." The concern is that the term "the general solution" would appear to imply that all solutions are special cases. There may in fact be many general solutions, but in most cases the difference between the general solutions would not be of significance to us. Also, for many differential equations, there are certain solutions that are not special cases of a general solution. These solutions are called *singular solutions* and cannot be obtained from a general solution by assigning appropriate values to the arbitrary constants. Those solutions that can be obtained by assigning particular values to the arbitrary constants are called *particular solutions*. If a general solution contains all solutions of the differential equation, we further refer to it as the *complete solution*. For most of the physical problems that we will encounter, the singular solutions will seldom satisfy the specified conditions of the problem and will, therefore, be disregarded.

Example 5.4. In the section on variables separable equations, we considered the linear differential equation

$$\frac{dC}{dt} = -k\left(2 - \frac{C}{5}\right)\left(16 - \frac{4C}{5}\right)$$

which had the general solution

$$C = 10\frac{\left[\bar{K}\exp\left(-\frac{8}{5}kt\right) - 2\right]}{\left[\bar{K}\exp\left(-\frac{8}{5}kt\right) - 1\right]}$$

The two solutions of the differential equation $C/5 = 2$ and $4C/5 = 16$, which cannot be obtained from the general solution given here by assigning finite values of the arbitrary constant $\bar{K}$, are called *singular solutions*.

Example 5.5. The reader can verify that a general solution of the nonlinear first order differential equation

$$\left(\frac{dC}{dx}\right)^2 + 2x\frac{dC}{dx} - 2C = 0$$

is

$$C = 2K(K + x)$$

A singular solution to the equation is

$$C = -\frac{1}{2}x^2$$

so called because it cannot be generated from the general solution $C = 2K(K + x)$ regardless of the value of the constant K. We have had occasion to solve for the value of a constant appearing in an equation by requiring the dependent variable to have a certain value for a specific value of the independent variable. For example, a condition requiring that $C = 2$ when $t = 0$ for the equation $C = Ke^{kt}$ allows us to solve for K—that is, setting $C = 2$ and $t = 0$ in the equation gives $2 = K$, so that the particular solution is given by $C = 2e^{kt}$. A specified condition similar to that above is commonly referred to as an *initial condition*. The total problem of the differential equation plus the initial condition is called an *initial value problem*. The number of conditions that can be satisfied in a problem is in general equal to the number of *independent or essential constants* appearing in the solution. An nth-order differential equation with a solution containing n arbitrary constants can satisfy n conditions. If the n conditions that the dependent variable is to satisfy are all specified for a single value of the independent variable, they are called *initial conditions*. For example, the solution to a second-order differential equation may have the following specified initial conditions

$$C(1) = 1, \qquad \frac{dC(1)}{dx} = 0, \qquad 1 \le x \le \infty$$

If the n conditions that the dependent variable is to satisfy are specified for extreme values of the independent variable, they are called *boundary conditions*. For example, a second-order differential equation may have the following boundary conditions

$$C(1) = 1, \qquad \frac{dC(2)}{dx} = 0, \qquad 1 \le x \le 2$$

The total problem of the differential equation plus the boundary conditions is called a *boundary value problem*.

The term *initial condition* seems particularly appropriate when time (t) is the independent variable and *initial* refers to $t = 0$; the term *boundary condition* seems most appropriate when a spatial variable such as x, y, or z is the independent variable, and *boundary* refers to a point on a physical boundary. The reader has probably encountered this common usage in his studies; however, in the following examples the specified conditions are identified in accordance with the definitions given here.

Other types of specified conditions such as

$$C(1) = C(0) + 2, \qquad C(1) = 2\,C(0), \qquad \frac{dC(1)}{dx} + C(1) = 0$$

may appear in problems formulated by the reader.

We stated earlier that the solution to an nth-order differential equation will in general possess n arbitrary constants. Since we have only had an opportunity at this point to solve a few first-order equations, the agreement of this statement with our experience cannot be adequately discussed. We state without proof that the solution to an nth-order differential equation will possess *no more* than *n essential* or *independent arbitrary constants*. However, there may exist cases in which the solution contains less than n independent arbitrary constants. We consider a set of arbitrary constants in an expression to be independent if they cannot be replaced with a smaller set by mathematical manipulation or rearrangement of the expression. Generally, a function of one or more arbitrary constants can be replaced by a single arbitrary constant. The following examples should illustrate this concept:

Example 5.6. A general solution to the differential equation $dC/dx = 0$ is $C = K_1/K_2$, where K_1 and K_2 are constants. Certainly this expression can be rearranged to reduce the number of constants to one simply by setting $K = K_1/K_2$. Either form, K or K_1/K_2 could only be used to satisfy one condition on C. For example, if $C(x)$ at $x = 0$ was known to be equal to 1—that is, $C(0) = 1$—then $K = K_1/K_2 = 1$. The particular solution to the differential equation that satisfied the initial condition is $C = 1$. It will sometimes be convenient to replace a constant K by a function of this constant. For example, replace K by $\ln K$ or by e^K, which are still constants.

Example 5.7. A general solution to the equation $dC/dx = 3C$ is $C = K_1e^{3x+K_2}$, where K_1 and K_2 are arbitrary constants. This expression can be rearranged as follows

$$C = K_1e^{3x+K_2} = K_1e^{3x}e^{K_2} = Ke^{3x}$$

and the solution that originally had two constants had been reduced to one, containing only one independent constant. As a check on these results we attempt to satisfy the *two* conditions $C(0) = 1$ and $C(1) = 2$. Using $C = Ke^{3x}$ the first of these two conditions yields $1 = K$; therefore, $C = e^{3x}$. The second condition cannot be satisfied since $C(1) = e^3 \neq 2$. If the solution in the form $C = K_1 e^{3x+K_2}$ contained two independent constants, both conditions could be satisfied. The conditions on C yield $1 = K_1 e^{K_2}$ and $2 = K_1 e^{3+K_2}$. The second expression can be written as $2 = K_1 e^{K_2} e^3$; but, since $K_1 e^{K_2} = 1$ from the first expression, this reduces to $2 = e^3$ which is not possible.

Example 5.8. The expression

$$C = K_1 \sin^2 t + K_2 \cos^2 t + K_3 \cos 2t$$

contains three arbitrary constants, but since $\cos 2t = \cos^2 t - \sin^2 t$, it can be written as $C = (K_1 - K_3) \sin^2 t + (K_2 + K_3) \cos^2 t = K_4 \sin^2 t + K_5 \cos^2 t$, where $K_4 = K_1 - K_3$ and $K_5 = K_2 + K_4$. This new expression has only two arbitrary constants, and this is the minimum number that this expression can possess. Therefore, the two arbitrary constants are independent.

We have seen that the solutions to the first-order differential equations solved in this chapter have contained only one independent constant. Even the two solutions to the first-order nonlinear equation given in example 5.8—$C = 2K(K + x)$ and $C = x^2/2$—can satisfy only one condition, since there is only one independent constant K. The following example will reveal a first-order differential equation whose solution appears to contain two independent constants.

Example 5.9. The differential equation

$$\frac{dC}{dx} = 3\frac{C}{x}$$

has a general solution

$$C = \begin{cases} K_1 x^3 & x \leq 0 \\ K_2 x^3 & x \geq 0 \end{cases}$$

Is this more general than the general solution $C = Kx^3$? We point out that for each range of the independent variable $x \leq 0$ and $x \geq 0$, the solution contains only one independent constant. However, since two conditions can be satisfied, the solution is in fact more general than $C = Kx^3$. For example, if the solution to the equation had to satisfy the conditions $C(-1) = 2$ and $C(1) = 1$, only the first form of the solution would work. We verify this by solving for K_1 and K_2 by applying two conditions on C to the general expression for C to get $K_1 = -2$ and $K_2 = 1$. Hence

$$C = \begin{cases} -2x^3 & x \leq 0 \\ x^3 & x \geq 0 \end{cases}$$

Example 5.9 reveals that in certain cases we have to generalize our statement regarding the number of arbitrary constants in a solution to allow for situations similar to the ones shown. We therefore state that no more than n arbitrary constants are contained in any segment of a solution to an nth-order differential equation. A segment of a solution is defined as the solution for a specific range of the independent variable.

Homogeneous Equations. A function $f(x,C)$ is called *homogeneous of degree n* if

$$f(1x,\ lC) = 1^n f(x,C) \tag{5.56}$$

Equivalently, if x and C are replaced by lx and lC in the function, l^n factors out, and becomes a multiple of the original equation.

Example 5.10.

$$f(x,C) = x^2 + 2xC + C^2; \qquad f(1x,lC) = 1^2(x^2 + 2xC + C^2)$$

$$f(x,C) = \sin\frac{C}{x} + \exp\left(\frac{C}{x}\right); \qquad f(1x,lC) = 1^0\left[\sin\frac{C}{x} + \exp\left(\frac{C}{x}\right)\right]$$

$$f(x,C) = \frac{\ln C/x}{C + 2x}; \qquad f(1x,lC) = 1^{-1}\left(\frac{\ln C/x}{C + 2x}\right)$$

$$f(x,C) = \exp\left(\frac{C}{x}\right) + \frac{\sqrt{(x + y)}}{x}; f(lx,lC) = \exp\left(\frac{C}{x}\right) + \frac{1}{\sqrt{1}} \times \frac{\sqrt{(x + y)}}{x}$$

The first three functions are homogenous of degree 2, 0, and -1, respectively, and the fourth function is not homogeneous.

The term *homogeneous* has already been used to describe a type of linear differential equation in earlier sections. The term *homogeneous* is also used to describe a special first-order differential equation on which a certain substitution reduces the equation to one in which the variables are separable. This duplicative usage of this same term is unfortunate, but of such standard use that we will also accept it.

The differential equation

$$M(x,C)\,dx + N(x,C)\,dC = 0 \tag{5.57}$$

is called homogeneous if $M(x,C)$ and $N(x,C)$ are homogeneous functions of the same degree.

If equation (5.57) is written as

$$\frac{dC}{dx} = -\frac{M(x,C)}{N(x,C)} \tag{5.58}$$

we see that the right-hand side will be a homogeneous function of degree zero—that is if $M(lx,lC) = l^n M(x,C)$, then $N(lx,lC) = l^n N(x,C)$, and therefore $M(lx,lC)/N(lx,lC) = M(x,C)/N(x,C)$, a homogeneous function of degree zero. Letting $l = x^{-1}$ we see that

$$-\frac{M(x,C)}{N(x,C)} = -\frac{M(lx,lC)}{N(lx,lC)} = -\frac{M(1,C/x)}{N(1,C/x)} = f\left(\frac{C}{x}\right) \tag{5.59}$$

where we have emphasized the fact that $M(1, C/x)/ N(1, C/x)$ is a function of the variable C/x by writing it as $f(C/x)$. In this case the equation can be written as

$$\frac{dC}{dx} = f\left(\frac{C}{x}\right) \tag{5.60}$$

which is considered a standard form for a homogeneous first-order differential equation. An alternate definition for a homogeneous differential equation then would be an equation in the form of equation (5.60) or one that can be put into this form.

If we make the substitution $C = vx$ into equation (5.60), then $dC/dx = v + x\,dv/dx$, and equation (5.60) can be written as

$$v + x\frac{dv}{dx} = f(v) \tag{5.61}$$

Assuming $f(v) \neq v$, this equation can be easily expressed as the variables separable equation

$$\frac{dv}{f(v) - v} = \frac{dx}{x} \tag{5.62}$$

whose solution has already been discussed. After the equation is integrated, we replace v by its value C/x.

If $f(v) = v$ the original differential equation [equation (5.60)] reduces to

$$\frac{dC}{dx} = \frac{C}{x} \tag{5.63}$$

an equation whose variables are readily separated.

Example 5.11. Suppose the concentration of dissolved organic matter (C) varies with distance in a stream according to

$$\frac{dC}{dx} = \frac{x^2 + C(2C + x)}{2Cx + x^2} = \frac{1 + \frac{C}{x}\left(2\frac{C}{x} + 1\right)}{2\frac{C}{x} + 1}$$

Solve the differential equation.

The right-hand side can be written as a function of C/x, so the differential equation is homogeneous in the form $dC/dx = f(C/x)$. Our treatment of this equation showed that the substitution $C = vx$ reduces the equation to the form of equation (5.62). Since $f(v) - v = 1/(2v + 1)$, we have

$$\frac{dv}{1/(2v + 1)} = \frac{dx}{x}$$

which can be integrated to yield

$$v^2 + v = \ln x + \ln K$$

or

$$v(v + 1) = \ln Kx$$

where, for convenience, we wrote $\ln K$ instead of K as an integration constant. Replacing v by C/x and simplifying, we get the final solution

$$C(C + x^2) - x^2 \ln Kx = 0$$

Exact Equations. The total derivative of a function was introduced in chapter 3. By investigating the change in a dependent variable or function due to incremental changes in the independent variables, we arrived at the *total differential* of the function $C(x,t)$ as

$$dC = \frac{\partial C}{\partial x}dx + \frac{\partial C}{\partial t}dt \tag{5.64}$$

The terms dx and dt are called the *differential of* x and the *differential of* t, respectively, and represent incremental changes in the independent variables.

We can analogously write the total differential of the function $U = U(x,C)$ as

$$dU = \frac{\partial U}{\partial x}dx + \frac{\partial U}{\partial C}dC \tag{5.65}$$

where U is the dependent variable, and x and C are independent variables. Equation (5.65) can also be termed an *exact* differential of the function U.

$$M(x,C)\,dx + N(x,C)\,dC = 0 \tag{5.66}$$

is *exact* if we can find a function $U(x,C)$ such that

$$dU\,(x,C) = M(x,C)\,dx + N(x,C)\,dC \tag{5.67}$$

This differential equation then takes the form

$$dU\,(x,C) = 0 \tag{5.68}$$

since the right-hand side has already been shown to be equal to 0, and the solution is simply

$$U(x,C) = K \tag{5.69}$$

By comparing equations (5.65) and (5.67), we see that for an exact differential equation

$$M = \frac{\partial U}{\partial x}, \qquad N = \frac{\partial U}{\partial C} \tag{5.70}$$

Differentiating the first of these equations with respect to C and the second with respect to x yields

$$\frac{\partial M}{\partial C} = \frac{\partial^2 U}{\partial x \partial C}, \qquad \frac{\partial N}{\partial x} = \frac{\partial^2 U}{\partial C \partial x}$$

For most of the functions that would be of interest to us, the order of differentiation is immaterial, so that

$$\frac{\partial^2 U}{\partial x \partial C} = \frac{\partial^2 U}{\partial C \partial x} \tag{5.71}$$

We therefore have

$$\frac{\partial M}{\partial C} = \frac{\partial N}{\partial x} \tag{5.72}$$

which is a condition that will insure that a differential equation of the form $M\,dx + N\,dC = 0$ is exact.

Familiarity with various common differential expressions will certainly facilitate the identification, and solution of exact differential equations. We list a few of these here:

$$x\,dx + C\,dC = d\left(\frac{x^2 + C^2}{2}\right)$$

$$x\,dC + C\,dx = d(xC)$$

$$\frac{x\,dC - C\,dx}{x^2} = d\left(\frac{C}{x}\right)$$

$$\frac{C\,dx - x\,dC}{C^2} = d\left(\frac{x}{C}\right)$$

$$\frac{x\,dx \pm y\,dy}{\sqrt{(x^2 \pm y^2)}} = d[\sqrt{(x^2 \pm y^2)}]$$

$$\frac{C\,dx + x\,dC}{Cx} = d(\ln Cx)$$

$$\frac{x\,dC - C\,dx}{Cx} = d\left(\ln\frac{C}{x}\right)$$

$$\frac{C\,dx - x\,dC}{Cx} = d\left(\ln\frac{x}{C}\right)$$

$$\frac{x\,dx + y\,dy}{x^2 + y^2} = d\left[\frac{1}{2}\ln(x^2 + y^2)\right]$$

$$ke^{kCx}(C\,dx + x\,dC) = de^{kCx}$$

$$ke^{kC/x}\left(\frac{x\,dC - C\,dx}{x^2}\right) = de^{kC/x}$$

$$\frac{x\,dC + C\,dx}{1 + C^2x^2} = d(\tan^{-1} Cx)$$

$$\frac{x\,dC - C\,dx}{x^2 + C^2} = d(\tan^{-1} C/x)$$

Although there are general mathematical techniques available to solve exact differential equations, we use a method of inspection, where we group terms of the equation into recognizable forms similar to those above.

Example 5.12. Solve the following differential equation:

$$\left(C + x + x^2C - \frac{1}{x} + e^{Cx}C - \frac{C}{x^2}\right)dx + \left(x + \frac{x^3}{3} + \frac{1}{C} + e^{Cx}x + \frac{1}{x}\right)dC = 0$$

We first test for exactness by using equation (5.72):

$$\frac{\partial M}{\partial C} = 1 + x^2 + e^{Cx} + Cxe^{Cx} - \frac{1}{x^2} = \frac{\partial N}{\partial x}$$

Since $\partial M/\partial C = \partial N/\partial x$, the differential equation is exact. We therefore group the terms of this equation as

$$(C\,dx + x\,dC) + \left(x^2C\,dx + \frac{x^3}{3}\,dC\right) + \left(-\frac{1}{x}\,dx + \frac{1}{C}\,dC\right)$$
$$+ e^{Cx}\,(C\,dx + x\,dC) + \left(\frac{1}{x}\,dC - \frac{C}{x^2}\,dx\right) = 0$$

or

$$d(Cx) + d\left(C\,\frac{x^3}{3}\right) + d\,(\ln C - \ln x) + d\left(e^{Cx}\right) + d\left(\frac{C}{x}\right) = 0$$

Therefore the solution is

$$Cx + \frac{Cx^3}{3} + \ln C/x + e^{Cx} + C/x = 0$$

The method of grouping terms by inspection is very quick if the invesitgator has the experience to readily identify various differentials. Those investigators encountering exact equations on a regular basis must therefore get this experience through practice or go to a more general method of solution.

Integrating Factors. When a differential equation $M\,dx + N\,dC = 0$ is not exact, it fails to satisfy equation (5.72)—that is, $\partial M/\partial C \neq \partial N/\partial x$. The equation can be made exact, however, through multiplication by a suitable function $p(x,C)$. The resulting exact differential equation

$$[p(x,C)][M(x,C)\,dx + N(x,C)\,dC] = 0 \tag{5.73}$$

will satisfy the following test for exactness

$$\frac{\partial(pM)}{\partial C} = \frac{\partial(pN)}{\partial x} \tag{5.74}$$

There are general methods available for determining integrating factors for various differential equations, but often it is possible to render a differential equation exact by inspection.

Example 5.13. Solve the differential equation

$$C(1 + x^2)\,dx + x\,dC = 0$$

Since $\partial M/\partial x = C \neq \partial N/\partial C = 0$, this equation is not exact. However, if we divide by Cx and group terms in the form

$$\frac{C\,dx + x\,dC}{Cx} + x\,dx = 0$$

which can be written as

$$d(\ln Cx) + d\frac{x^2}{2} = 0$$

Integration then gives the solution

$$\ln Cx + \frac{x^2}{2} + K = 0$$

Thus for this differential equation the integrating factor was $p(x,C) = 1/Cx$.

In this last example the method of separation of variables would also have been applicable. For example, if we divide the equation by Cx we get

$$\left(\frac{1}{x} + x\right) dx + \frac{dC}{C} = 0$$

which can be integrated term by term to yield

$$\ln x + \frac{x^2}{2} + \ln C + K = 0$$

or

$$\ln Cx + \frac{x^2}{2} + K = 0$$

which is the same answer determined using an integrating factor. This equation is of the form $M\,dx + N\,dC = 0$, where $M = (1/x) + x$ and $N = C^{-1}$, and the equation is exact since $\partial M/\partial C = 0 = \partial N/\partial x$. Note that the variables separable differential equation

$$M(x)\,dx + N(C)\,dC = 0$$

is always an exact equation since $\partial M(x)\,\partial C = \partial N(C)/\partial x = 0$. We see that the process of separating the variables in a suitable differential equation is equivalent to converting an equation that is not exact into one that is by the use of an integrating factor.

Example 5.14. Solve the differential equation

$$(C + xe^{-Cx})\,dx + x\,dC = 0$$

We see that since $\partial M/\partial C = 1 - x^2e^{-Cx} \neq \partial N/\partial x = 1$, the equation is not exact. It can also be seen that the equation is not separable. However, if we multiply the equation by e^{Cx} we get

$$(Ce^{Cx} + x)\,dx + xe^{Cx}\,dC = 0$$

and now $\partial M/\partial C = Cxe^{Cx} + e^{Cx} = \partial N/\partial x$. The equation is now exact, and we call the term e^{Cx} an integrating factor for the differential equation.

The equation can be rearranged in the form

$$e^{Cx}(C\,dx + x\,dC) + x\,dx = 0$$

or

$$d(e^{Cx}) + d\left(\frac{x^2}{2}\right) = 0$$

Integration yields the solution

$$e^{Cx} + \frac{1}{2}x^2 + K = 0$$

Linear Equations. A linear differential equation is one that can be written in the form

$$\frac{dC}{dx} + P(x)C = Q(x) \tag{5.75}$$

which was previously introduced in the section on homogeneous linear equations. A property of the linear differential equations of major importance is the *superposition of solutions*. Suppose the response (solution) of a particular system for an input $Q_1(x)$ is known to be $C_1(x)$. Similarly, the response for input $Q_2(x)$ is $C_2(x)$. Then by superposition, the response of the system for the input $Q_3(x) = Q_1(x) + Q_2(x)$ is simply $C_3(x) = C_1(x) + C_2(x)$. For example, if the solution for $Q_1(x)$ is $C_1(x)$, the solution to $2Q_1(x)$ is $2C_1(x)$, that is, if the input is doubled the response is doubled. The superposition of solutions is therefore a most important characteristic of linear differential equations as well as of linear partial differential equations.

It has been shown previously that equation (5.75) has $p(x) = \exp(\int p dx)$ as an integrating factor which involves only one variable. To verify this, we multiply both sides of (5.75) by this factor and get

$$p(x)\frac{dC}{dx} + p(x)C\,p(x) = Q(x)\,p(x) \tag{5.76}$$

The solution of this equation is just the derivative of the product $p(x)C$, so we write the equation in the equivalent form

$$\frac{d}{dx}[p(x)C] = Q(x)\,p(x) \tag{5.77}$$

Integrating this equation, we obtain

$$p(x)C = \int Q(x)\,p(x)\,dx + K \tag{5.78}$$

Dividing by $p(x)$ and noting that $1/p(x) = \exp(-\int p dx)$, we have

$$C = \exp\left(-\int p\,dx\right) \int Q(x) \exp\left(\int p\,dx\right) dx + K\exp\left(-\int p\,dx\right) \qquad (5.79)$$

We did not include the linear differential equation as a special case in the section on integrating factors because of its importance in modeling many practical environmental problems; moreover, because of its simplicity of solution, we treat it here as a separate topic.

An investigator, when confronted by a linear differential equation, has the option of attempting to find an integrating factor directly or making use of equation (5.79). We note here that equation (5.79) is valid for the homogeneous equation where $Q(x) = 0$.

Example 5.15. Solve the equation

$$\frac{dC}{dx} + \frac{1}{x}C = 6$$

To use equation (5.79), we need the integral

$$\exp\left(\int p\,dx\right) = \exp\left(\int \frac{1}{x}\,dx\right) = \exp\left(\ln x\right) = x$$

Substituting into equation (5.79), we have

$$C = \frac{1}{x} \int 6x\,dx + K\frac{1}{x}$$

The general solution is therefore

$$C = 3x + \frac{K}{x}$$

Another approach to the solution of this equation would be the method described in the previous section. To this end we write the equation in the alternate form

$$dC + \frac{C}{x}\,dx = 6\,dx$$

obtained by multiplying both sides by dx. The test for exactness reveals that this equation is not exact but can be made exact by multiplying both sides by the integrating factor x to obtain

$$x\,dC + C\,dx = 6x\,dx$$

This can be written by inspection as

$$d(xC) = d(3x^2)$$

which can be integrated to yield

$$xC = 3x^2 + K$$

or

$$C = 3x + \frac{K}{x}$$

We had to guess the integrating factor $p(x) = x$ for this method, whereas our development in this section gives it to us by use of the equation

$$p(x) = \exp\left(\int p dx\right) = x$$

The reader should be able to see that for linear differential equations, the use of equation (5.79) will in general be more suitable than the methods of previous sections.

Example 5.16. Solve the equation

$$\frac{dC}{dx} + \frac{1}{x}C = e^{-x}$$

From the last example we know $\exp(\int p dx) = x$, therefore substitution into equation (5.79) yields

$$C = \frac{1}{x} \int e^{-x} x \, dx + \frac{K}{x}$$

Evaluating the integral as

$$\int xe^{-x} \, dx = -e^{-x}(x + 1)$$

the general solution is seen to be

$$C = -\left(\frac{x + 1}{x}\right)e^{-x} + \frac{K}{x}$$

The reader should verify that this is a solution by substitution into the original differential equation.

Bernoulli's Equation. A slightly more general differential equation than the linear equation (5.75) is the Bernoulli equation

$$\frac{dC}{dx} + p(x)C = Q(x)C^n \qquad (n \neq 1) \tag{5.80}$$

where n is a real number. This differential equation is nonlinear for all values of n other than 0 or 1. For $n = 0$, equation (5.80) reduces to the linear form of equation (5.75). For $n = 1$, equation (5.80) can be written as

$$\frac{dC}{dx} + R(x)C = 0 \tag{5.81}$$

where $R(x) = p(x) - Q(x)$, therefore equation (5.81) is a homogeneous linear differential equation.

For cases other than $n = 0$ and 1, we write equation (5.80) in the form

$$C^{-n}\frac{dC}{dx} + P(x)C^{1-n} = Q(x) \tag{5.82}$$

The solution $C = 0$ for $n > 0$ in (5.80) is precluded by the form (5.82). This equation will be linear in the variable u if we use the transformation

$$u = C^{1-n} \qquad (C \neq 0)$$

Note that

$$\frac{du}{dx} = (1 - n)C^{-n}\frac{dC}{dx} \tag{5.83}$$

and therefore (5.82) becomes

$$\frac{1}{1-n} \times \frac{du}{dx} + P(x)u = Q(x) \tag{5.84}$$

which is a linear first-order equation for the variable u. After solving the equation for u by methods previously introduced for linear differential equations, the solution to (5.80) is given by $C = u^1/(1 - n)$.

Example 5.17. Solve the equation

$$\frac{dC}{dx} - C = xC^2 \tag{5.85}$$

This is a special case of (5.80) with $P(x) = -1$, $Q(x) = x$, and $n = 2$. We write the equation as

$$C^2\frac{dC}{dx} - C^{-1} = x$$

and use the transformation $u = C^{-1}$ to obtain

$$-\frac{du}{dx} - u = x$$

To use the method from the previous section on linear differential equations, we write the equation as

$$\frac{du}{dx} + u = -x$$

and note that $p(x) = 1$ for the equation in this form. Then the solution form given by equation (5.79) yields

$$u = e^{-x} \int (-x)e^{x}\, dx + Ke^{-x}$$

Since

$$\int xe^{x}\, dx = e^{x}(x - 1)$$

then

$$u = -(x - 1) + Ke^{-x}$$

The solutions to our original differential equation are therefore

$$C = u^{-1} = (1 - x + Ke^{-x})^{-1}$$

and $C = 0$.

Example 5.18. Solve the equation

$$\frac{dC}{dx} + \frac{1}{x}C = \frac{k^2}{x^2}C^{-2}$$

We write the equation as

$$C^2\frac{dC}{dx} + \frac{1}{x}C^3 = \frac{k^2}{x^2}$$

and use the transformation $u = C^3$ to obtain

$$\frac{du}{dx} + \frac{3}{x}u = \frac{3k^2}{x^2}$$

Equation (5.79) then yields

$$u = x^{-3} \int \frac{3k^2}{x^2}x^3\, dx + Kx^{-3}$$

where

$$\exp\left(\int \frac{3}{x}dx\right) = \exp(3\ln x) = x^3$$

$$u = \frac{3k^2}{2x} + Kx^{-3}$$

therefore

$$C = \left(\frac{3k^2}{2x} + Kx^{-3}\right)^{1/3}$$

Remarks on Solution Techniques. We have discussed the various solution methods to first-order differential equations that should be most useful with common environmental needs. When the invesitgator encounters a differential equation, the first step in its treatment is the "keying out" or identifying step. If the equation is keyed out as one of those discussed in this chapter, the solution method explained can be utilized to obtain the appropriate solution. There may be a differential equation that falls into one of the categories discussed but yields integrals difficult to evaluate. We have decided to present the integration of only simple functions in our examples so as to limit the scope of the text.

The investigator's goal is to solve a given differential equation, and any method including (as we have previously mentioned) guessing is to be considered legitimate in reaching this goal. If the equation in question is keyed out as one whose solution we did not consider and guessing or direct observation fails, then the investigator should try various transformations or substitutions of variables so as to convert the equation to one whose form is more tractable. Recall, for example, the substitutions used in the methods of homogeneous equations, integrating factors, and Bernoulli's equation. Whatever technique is employed, the resulting solution should always be "verified" by substitution back into the original differential equation.

We conclude our treatment of first-order differential equations with a series of examples illustrating the solution process, beginning with the keying out of the equation type and ending with the verification of solution. The examples will also address the models developed in chapter 4.

Example 5.19. A representative steady-state dissolved oxygen model introduced in chapter 4 was

$$\frac{Q}{A} \times \frac{dC}{dx} = k_2(C_s - C) - k_1L - S_r$$

Since all quantities in this equation are constant with respect to the variable x, we write the equation in the form

$$\frac{dC}{dx} + K_1C = K_2$$

where $K_1 = Ak_2/Q$ and $K_2 = A(k_2C_s - k_1L - S_r)/Q$. We recognize this equation as an ordinary (no partial derivations), first-order (highest order derivative is first order), first-degree (exponent of highest order derivative equals 1), linear (exponent of C and its derivatives equals 1), nonhomogeneous (due to the term K_2, which is not of the first degree in C or its derivatives) differential equation.

The investigator may recognize the familiar left-hand side of this equation—that is, for the case of $K_2 = 0$, the resulting equation should be easily recognizable as one with the exponential function as its solution. Thus $C = Ke^{-K_1x}$ would satisfy

$$\frac{dC}{dx} + K_1C = 0$$

where K is an arbitrary constant. However, since the solution of the differential equation is not 0, we must search for a different solution. The clever investigator will realize that we have only to add an extra simple term to $C = Ke^{-K_1x}$ such as K_2/K_1 to get the solution to equal K_2 instead of 0. Substituting $C = Ke^{-K_1x} + K_2/K_1$ back into the differential equation yields an identity; thus we have found a general solution.

Suppose we were not clever enough to guess the term K_2/K_1 in example 5.19. We would next have to search through our solution techniques to find a suitable approach. Algebraic manipulations would reveal that the variables could not be separated because of the nonhomogeneous term K_2 in the equation. Because we have already keyed the equation out as linear, we try the solution method appropriate to linear differential equations. Our equation is certainly in the general form of equation (5.75): $dC/dx + P(x)C = Q(x)$, where in our case $P(x) = K_1$ and $Q(x) = K_2$. Therefore, substitution into equation (5.79) yields

$$C = \exp\left(-\int K_1\,dx\right) \int K_2\exp\left(\int K_1\,dx\right)dx + K\exp\left(-\int K_1\,dx\right)$$

Evaluating the integrals and simplifying gives us the solution

$$C = K_2/K_1 + Ke^{-K_1x}$$

which is what was previously obtained.

Substituting the values of the constants K_1 and K_2 yields

$$C = (k_2C_s - k_1L - S_r)k_2 + Ke^{-Ak_2x/Q}$$

where K is an arbitrary constant to be evaluated when a boundary condition is applied to C.

Example 5.20. Another model introduced in chapter 4 related the increase in biomass with time in a CFSTR operating with no biomaterial washout. The differential equation was

$$V\frac{dC}{dt} = QC_0 - QC + (R_g - R_d)M_o\exp(k_gR_g - k_dR_d)t$$

we write this equation in the simpler form

$$\frac{dC}{dt} + K_1C = K_2 + K_3e^{K_4t}$$

where the constants are defined to be $K_1 = Q/V$, $K_2 = QC_0/V$, $K_3 = (R_g - R_d)M_0/V$, and $K_4 = (k_gR_g - k_dR_d)$.

This equation, as was the differential equation in example 5.19, is ordinary, first order, first degree, linear, and nonhomogeneous. We therefore proceed directly to equation (5.79) for the solution and replace the variable x in that equation with the variable t. First $P(t) = K_1$, and $Q(t) = K_2 + K_3e^{K_4t}$, and $\exp(-\int pdt) = e^{-K_1t}$, so that equation (5.79) yields

$$C = e^{-K_1t}\int (K_2 + K_3e^{K_4t})e^{K_1t}\,dt + Ke^{-K_1t}$$

Evaluating the integrals $\int K_2e^{K_1t}\,dt$ and $\int K_3e^{K_4t}e^{K_1t}\,dt$ as $(K_2/K_1)e^{K_1t}$ and $[K_3/(K_4 + K_1)]e^{K_1t}$, respectively, we obtain a general solution

$$C = (K_2/K_1) + [K_3/(K_4 + K_1)] + Ke^{-K_1t}$$

Substituting the values of the constants K_1 through K_4 into the equations for C gives

$$C = C_0 + \frac{(R_g - R_d)M_0}{(k_gR_g - k_dR_d)V + Q} + Ke^{-Qt/V}$$

where K is an arbitrary constant.

The predictive model for microbial die-off developed in chapter 4 could be treated by methods similar to those described here for certain special cases. For example, if the model was integrated over the depth z, and if either the velocity terms V_x, V_y, and V_z were equal to 0, or C was independent of position, we would have an ordinary, first-order, first-degree, linear, *homogeneous* equation of the form

$$\frac{dC}{dt} + K_1C = 0$$

The solution would be immediately recognizable as an exponential function of t—that is, $C = Ke^{-K_1t}$. We would have a similar situation for the steady-state case if we considered only one component of the velocity. The resulting equation would be of the form

$$\frac{dC}{dx} + P(x)C = 0$$

where C is a function of the variable x. Since this equation is ordinary, first order, first degree, and linear, we could again make use of equation (5.79) to determine a solution. We may, however, run into trouble with the integrals depending on the complexity of the function $P(x)$.

Example 5.21. In the differential equation in example (5.20) the $P(x)$ term was derived from the multiplication of a constant term and $V_{\bar{x}}^{-1}$. If this product term was of the form $P(x) = K_1x^n$, where K_1 is a constant and n is a positive integer, the equation becomes $dC/dx + K_1x^nC = 0$. Making use of equation (5.79), we first evaluate

$$\exp\left(\int K_1x^n\,dx\right) = \exp\left(K_1\frac{x^{n+1}}{n+1}\right)$$

and therefore

$$C = K\exp\left(-K_1\frac{x^{n+1}}{n+1}\right)$$

Example 5.22. Solve the differential equation

$$x\frac{dC}{dx} + C\ln x = C\ln C + C$$

We key this equation out as ordinary, first order, and nonlinear ($\ln C$). We observe that the variables are not separable in this equation in its present form, and rewrite the equation in the alternate form

$$C[(\ln C - \ln x) + 1]dx - x\,dC = 0$$

Dividing by x, we get

$$C/x(\ln C/x + 1)\,dx - dC = 0 \qquad (x \neq 0)$$

Since this equation can be put in the form of equation (5.60),

$$\frac{dC}{dx} = f\left(\frac{C}{x}\right) = \frac{C}{x}\left(\ln\frac{C}{x} + 1\right)$$

it is a homogeneous equation.

We make the substitution $C = vx$, note that $dC/dx = v + x(dv/dx)$, and write the equation as

$$v + x\frac{dv}{dx} = v(\ln v + 1)$$

Simplifying and separating the variables, we get

$$\frac{dv}{v\ln v} = \frac{dx}{x}$$

Integration yields a general solution

$$\ln(\ln v) = \ln x + \bar{K}$$

or

$$\ln v = \exp(\ln x + \bar{K}) = Kx$$

where $K = e^{\bar{K}}$. Finally this can be written as

$$v = e^{Kx}$$

This determines C as

$$C = vx = xe^{Kx}$$

To verify this solution, we substitute back into the original differential equation, getting

$$x(e^{Kx} + xKe^{Kx}) + xe^{Kx} \ln x = xe^{Kx}(\ln x + Kx) + xe^{Kx}$$

which can be seen to be an identity.

Example 5.23. Solve the differential equation

$$(2x^2 - Ce^x)\,dx - e^x\,dC = 0$$

This equation is keyed out as ordinary, first order, first degree, linear, and nonhomogeneous. An alternate form is

$$\frac{dC}{dx} + C = 2x^2e^{-x}$$

This could be solved directly by the use of equation (5.79).

Example 5.24. Solve the differential equation

$$x\,\frac{dC}{dx} = x^3 + 2C$$

or

$$\frac{dC}{dx} - \frac{2}{x}C = x^2$$

This equation is keyed out as ordinary, first order, first degree, linear, and nonhomogeneous. It too can be solved by the use of equation (5.79).

Example 5.25. Solve the differential equation

$$2x\,\frac{dC}{dx} + x^2 + C^2 = 0$$

This equation is keyed out as ordinary, first order, first degree, nonlinear, and nonhomogeneous. An alternate form is

$$(x^2 + C^2)\,dx + 2xC\,dC = 0$$

Since $\partial(x^2 + C^2)/\partial C = 2C = \partial(2xC)/\partial x = 2C$, this equation is exact. The first term $x^2\,dx$ can by observation be written as $d(x^2/3)$. The remaining two terms are $C^2 dx + 2xC\,dC$. Comparison with our list of common differential expressions in the section on exact equations does not yield an immediate solution, so we add to that list the expression

$$d(x^m C^m) = mx^{m-1}C^n\,dx + nC^{n-1}x^m\,dC$$

For the values $m = 1$, $n = 2$, this becomes

$$d(xC^2) = C^2\,dx + 2Cx\,dC$$

Thus we can write our equation as

$$d\left(\frac{x^3}{3} + xC^2\right) = 0$$

Integration yields a general solution

$$x^3 + 3xC^2 = K$$

For a more direct solution of this equation, we use the expressions developed in the section on exact differential equations—that is, using

$$\frac{\partial U}{\partial x} = M, \qquad \frac{\partial U}{\partial C} = N$$

For our equation these expressions yield

$$\frac{\partial U}{\partial x} = x^2 + C^2, \qquad \frac{\partial U}{\partial C} = 2xC$$

Since $dU = M\,dx + N\,dC$, to arrive at a solution to our equation we have only to solve for U. We first determine U by integration of the partial derivative of U with respect to x, keeping C constant. In doing this we keep in mind that our arbitrary constant in this case would as usual not depend on x, but it may depend on C—that is, for integration when one variable is involved, we have

$$\frac{dU}{dx} = f(x), \qquad U = \int f(x)\,dx + K$$

However, when two variables are involved, we have

$$\frac{\partial U}{\partial x} = f(x,C), \qquad U = \int f(x,C)\,\partial x + g(C)$$

where the symbol ∂x denotes that integration is with respect to x only, and

$g(C)$ is our arbitrary constant for this type of integration. Thus for $\partial U/\partial x = x^2 + C^2$, we get

$$U = \int (x^2 + C^2)\,\partial x + g(c) = \frac{x^3}{3} + C^2x + g(C)$$

To determine $g(C)$, substitute this form of U back into the expression $\partial U/\partial C = 2xC$ to get

$$2Cx + \frac{\partial g(C)}{\partial C} = 2Cx$$

or

$$\frac{\partial g(C)}{\partial C} = 0$$

which yields

$$g(C) = K_1$$

Thus we have

$$U = \frac{x^3}{3} + C^2x + K_1$$

Recalling that $dU = M\,dx + N\,dc$, our original equation can be written as

$$dU = 0$$

so that a solution is

$$U = K_2$$

or

$$\frac{x^3}{3} + C^2x + K_1 = K_2$$

This is written as

$$x^3 + 3xC^2 = K$$

where $K = 3(K_2 - K_1)$ for simplicity and K_1 and K_2 were arbitrary constants. This is the same as the solution we obtained previously. This method is more direct, but the clever analyst will often find the previous method of inspection quicker.

Example 5.26. Solve the equation

$$\left(\frac{dC}{dx}\right)^2 + (3C - 2x)\frac{dC}{dx} - 6xC = 0$$

This differential equation is keyed out as ordinary, first order, second degree, and nonlinear. It can be factored algebraically to the form

$$\left(\frac{dC}{dx} - 2x\right)\left(\frac{dC}{dx} + 3C\right) = 0$$

Since either term can be equal to 0, we solve for both terms equal to 0. Separating the variables for these two differential equations, we have

$$dC = 2x\,dx, \qquad \frac{dC}{C} = -3\,dx$$

Integration yields the solution

$$(C - x^2 + K_1)(C - K_2e^{-3x}) = 0$$

where K_1 and K_2 are arbitrary constants.

Equations of Higher Order

In this section we consider ordinary differential equations that contain derivatives of the independent variable of order two or greater. In keeping within the scope of the book, we limit our treatment primarily to higher order linear differential equations. We treat here certain higher order equations that can be solved by techniques similar to those used for first-order equations.

Equations Solved by Direct Integration. The first differential equation discussed in this chapter was of the form

$$\frac{dC}{dx} = f(x) \tag{5.86}$$

which could be directly integrated to yield the solution

$$C = \int f(x)\,dx + K \tag{5.87}$$

The analogous higher order differential equation is

$$\frac{d^nC}{dx^n} = f(x)$$

where n is a positive integer. This equation must be integrated n times to yield a solution. The following examples will illustrate this technique.

Example 5.27. Consider the differential equation above for the case $n = 2$ and $f(x) = x^2$:

$$\frac{d^2C}{dx^2} = x^2$$

Integrating once yields

$$\frac{dC}{dx} = \frac{x^3}{3} + K_1$$

We integrate again to obtain the solution

$$C = \frac{x^4}{12} + K_1x + K_2$$

Note that we have produced two arbitrary constants K_1 and K_2 with the solution of the second-order equation. As we have previously mentioned, the solution to the nth-order differential equation will generally contain n arbitrary constants.

Example 5.28. Solve the equation

$$\frac{d^3C}{dx^3} = e^{2x} - \frac{1}{x^2}$$

Integrating three times successively yields

(1) $$\frac{d^2C}{dx^2} = \frac{1}{2}e^{2x} + \frac{1}{x} + K_1$$

(2) $$\frac{dC}{dx} = \frac{1}{4}e^{2x} + \ln x + K_1x + K_2$$

(3) $$C = \frac{1}{8}e^{2x} + x \ln x - x + K_1 \frac{x^2}{2} + K_2x + K_3$$

Our solution contains three arbitrary constants K_1, K_2, and K_3.

Second-Order Equations with the Dependent Variable Missing. A special differential equation of this type is shown here:

$$\frac{d^2C}{dx^2} + P(x)\frac{dC}{dx} = Q(x) \tag{5.88}$$

where it is seen that only derivatives of the dependent variable C, but not the variable itself, occur in the equation. An equation of this type is solved by a substitution of variables, which reduces the order of the equation.

Example 5.29. Solve the differential equation

$$\frac{d^2C}{dx^2} + \frac{1}{x} \times \frac{dC}{dx} = 2$$

We use the substitution

$$v = \frac{dC}{dx}$$

to reduce the equation to the form

$$\frac{dv}{dx} + \frac{1}{x} v = 0$$

which is immediately identifiable as a first-order, linear equation readily solved by methods already introduced. We could solve it directly using equation (5.79), but instead we observe that the equation can be written as

$$\frac{1}{x} \times \frac{d}{dx} (xv) = 0$$

which, upon integration gives

$$xv = K_1$$

Replacing v by dC/dx, we have

$$x(dC/dx) = K_1$$

or

$$\frac{dC}{dx} = K_1 x$$

This is directly integrated to give

$$C = K_1 \frac{x^2}{2} + K_2$$

which contains two arbitrary constants.

Second-Order Equations with the Independent Variable Missing. A differential equation of this type is

$$\frac{d^2C}{dx^2} + P(C) \frac{dC}{dx} = Q(C)$$

where it is seen that there is no function of the independent variable x appearing in the equation except C itself. This equation can be reduced to a first-order equation by the same substitution—that is, $dC/dx = v$.

Example 5.30. Solve the differential equation

$$C \frac{d^2C}{dx^2} - \left(\frac{dC}{dx}\right)^2 = 3$$

Due to the presence of C in the equation, we write

$$\frac{d^2C}{dx^2} = \frac{dv}{dx} = \frac{dv}{dC} \times \frac{dC}{dx} = \frac{dv}{dC} v$$

Substituting back into the equation, we get

$$Cv \frac{dv}{dC} - v^2 = 3$$

The variables can be separated to yield

$$\frac{v\,dv}{3 + v^2} = \frac{dC}{C}$$

Integration then gives

$$\tfrac{1}{2} \ln (v^2 + 3) = \ln C + \ln K_1$$

where we have written $\ln K_1$ instead of K_1 for convenience. Thus

$$v^2 + 3 = (K_1C)^2$$

or

$$v = \frac{dC}{dx} = \sqrt{[(K_1C)^2 - 3]}$$

and

$$\frac{dC}{\sqrt{[(K_1C)^2 - 3]}} = dx$$

Integration yields the solution

$$\frac{1}{K_1} \ln [C + \sqrt{(C^2 - 3/K_1^2)}] = x + K_2$$

or

$$C + \sqrt{(C^2 - 3/K_1^2)} = K_2 e^{K_1 x}$$

Example 5.31. Solve the differential equation

$$\frac{d^2C}{dx^2} + C = 0$$

Again using the substitution $v = dC/dx$, we get

$$v \frac{dv}{dC} + C = 0$$

The variables can be separated to yield

$$v\,dv = -C\,dC$$

which upon integration yields

$$\frac{v^2}{2} = -\frac{C^2}{2} + \frac{K_1}{2}$$

where we have written $K_1/2$ instead of K_1 for convenience. Solving for v and setting $v = dC/dx$ gives

$$\frac{dC}{\sqrt{(K_1 - C^2)}} = dx$$

Integration yields the solution

$$\sin^{-1}(C/K_1) = x + K_2$$

or

$$C = K_1 \sin(x + K_2)$$

Linear Differential Equations

We have previously introduced the general first-order linear differential equation as

$$\frac{dC}{dx} + P(x)C = Q(x) \tag{5.89}$$

The general second-order linear differential equation can be written as

$$\frac{d^2C}{dx^2} + P(x)\frac{dC}{dx} + Q(x)C = R(x) \tag{5.90}$$

Similarly, the general nth-order linear differential equation can be written as

$$\frac{d^nC}{dx^n} + P_1(x)\frac{d^{n-1}C}{dx^{n-1}} + \ldots + P_{n-1}(x)\frac{dC}{dx} + P_n(x)C = R(x) \tag{5.91}$$

For the general case, when the coefficients $P_1, P_2, \ldots P_n$ are functions of x, the treatment of the differential equation becomes complex and beyond the scope of this text. We therefore concentrate our efforts on higher order linear equations with *constant coefficients*.

Linear Differential Equations of the Second Order. In this section we will introduce terminology and techniques within the context of second-order equations, which can in general be applied to equations of higher order.

The general second-order differential equation is called *homogeneous* if $R(x) = 0$ and *nonhomogeneous* otherwise. The homogeneous equation

$$\frac{d^2C}{dx^2} + P\frac{dC}{dx} + QC = 0 \tag{5.92}$$

is referred to as the *complementary* or *reduced* equation. The solution to this equation is then called a *complementary solution* or *integral*, and will in general contain two arbitrary constants. Any solution to equation (5.90), no matter how special, will be called a *particular solution* or *integral* and contains no arbitrary constants.

General Solution. The general solution to equation (5.90) is defined as the sum of a particular solution (C_p) to this equation, and the complementary solution (C_c). Thus we write

$$C = C_p + C_c \tag{5.93}$$

where $C =$ general solution.

Superposition of Solutions. If $C_1(x)$ and $C_2(x)$ are solutions to the reduced equation (5.92), then any linear combination of these solutions is also a solution—that is,

$$C = K_1\, C_1(x) + K_2\, C_2(x) \tag{5.94}$$

where K_1 and K_2 are arbitrary constants, is also a solution. We can verify this by substitution into equation (5.92):

$$\frac{d^2}{dx^2}[K_1C_1 + K_2C_2] + P\frac{d}{dx}[K_1C_1 + K_2C_2] + Q\,[K_1C_1 + K_2C_2] = 0 \tag{5.95}$$

Carrying out the differentiation and simplifying

$$K_1\left(\frac{d^2C_1}{dx^2} + P\frac{dC_1}{dx} + QC_1\right) + K_2\left(\frac{d^2C_2}{dx^2} + P\frac{dC_2}{dx} + QC_2\right) = 0 \tag{5.96}$$

Since C_1 and C_2 are solutions, both brackets vanish identically.

The Complementary Second-Order Equation with Constant Coefficients. The complementary equation for the general first-order linear differential equation is obtained by setting $Q(x) = 0$ in equation (5.89). This yields

$$\frac{dC}{dx} + P(x)C = 0 \qquad (5.97)$$

If we consider the case where P is a constant, then the complete solution to equation (5.97) can be obtained as

$$C = Ke^{-Px} \qquad (5.98)$$

Note that we could have assumed a solution of the form $C = e^{mx}$, where m *is* a constant substituted in equation (5.97) to obtain the equation

$$(m + P)e^{mx} = 0 \qquad (5.99)$$

Thus $m = -P$, and since e^{mx} can never vanish, our solution is

$$C = e^{-Px} \qquad (5.100)$$

We arrive at the same solution as before simply by multiplying this solution by an arbitrary constant K to get

$$C = Ke^{-Px} \qquad (5.101)$$

We now try to determine if a similar technique will work for the second-order equation with constant coefficients. Substituting $C = e^{mx}$ in equation (5.92) yields $(m^2 + Pm + Q)e^{mx} = 0$. Since e^{mx} cannot vanish, we have the equation

$$(m^2 + Pm + Q) = 0 \qquad (5.102)$$

This equation is called the *characteristic* or *auxiliary equation* of equation (5.92) and is a quadratic equation in m. Its solutions m_1 and m_2 will give the two solutions to the differential equation $C_1 - e^{m_1x}$, and $C_2 = e^{m_2x}$. Since we expect the second-order equation to have two arbitrary constants, the complementary solution, using superposition, would be

$$C = K_1e^{m_1x} + K_2e^{m_2x} \qquad (5.103)$$

Example 5.32. Solve the differential equation

$$\frac{d^2C}{dx^2} + 5\frac{dC}{dx} - 6C = 0$$

The auxiliary equation is

$$(m^2 + 5m - 6) = 0$$

The two solutions are $m_1 = -2$ and $m_2 = -3$, thus the complementary solution is

$$C = K_1 e^{-2x} + K_2 e^{-3x}$$

which can be easily verified by substitution.

The solutions to the auxiliary equation (5.102) are

$$m = -\frac{P}{2} \pm \frac{1}{2}\sqrt{(P^2 - 4Q)} \tag{5.104}$$

We can see that the complementary solution $C = K_1 e^{m_1 x} + K_2 e^{m_2 x}$ will take on very different forms depending on whether the discriminant $(P^2 - 4Q)$ is positive, zero, or negative. The nature of the roots m_1 and m_2 for these cases is as follows:

(A) $P^2 - 4Q > 0$. The roots are real and unequal.

(B) $P^2 - 4Q = 0$. The roots are real and equal.

(C) $P^2 - 4Q < 0$. The roots are complex and unequal.

(A) For this case we write

$$m_1 = -\frac{P}{2} + \frac{1}{2}\sqrt{(P^2 - 4Q)}$$

$$m_2 = -\frac{P}{2} - \frac{1}{2}\sqrt{(P^2 - 4Q)}$$

and we have the two real unequal solutions, $e^{m_1 x}$ and $e^{m_2 x}$ which can be superposed to yield the complementary solution

$$C = K_1 e^{m_1 x} + K_2 e^{m_2 x} \tag{5.105}$$

An example of this type of solution was given earlier.

(B) For equal roots to occur the discriminant must vanish $P^2 = 4Q$, and $m_1 = m_2 = m = P/2$. In this case the superposition of solutions would yield

$$C = K_1 e^{mx} + K_2 e^{mx} = (K_1 + K_2)e^{mx} = Ke^{mx} \tag{5.106}$$

a solution with only one arbitrary constant.

To find a second solution we use a fundamental concept of differential equations—that is, if we know a solution $C = C_1$ of a second-order differential equation, then the substitution $C = C_1 v$ will reduce the equation to one of the first order in dv/dx. Substituting $C = C_1 v$ in equation (5.92) yields

$$\frac{d^2 v}{dx^2} = 0 \tag{5.107}$$

which can be integrated directly as $v = K_1x + K_2$. Thus our complementary solution for this case is

$$C = (K_1x + K_2)e^{mx} \tag{5.108}$$

To get the second solution for the case of equal roots, we have only to multiply the first solution by x. This concept can also be extended to equations of higher order having equal roots.

(C) For this case, we write $P^2 - 4Q$ (which is < 0) as $-(4Q - P^2)$, so that the two roots can be written as

$$m_1 = -\frac{P}{2} + \frac{1}{2}i\sqrt{(4Q - P^2)}$$

$$m_2 = -\frac{P}{2} - \frac{1}{2}i\sqrt{(4Q - P^2)}$$

where we have used the relation $i = \sqrt{-1}$. For convenience, let $a = P/2$ and $b = \sqrt{(4Q - P^2)}/2$. Thus the roots can be written as

$$m_1 = a + ib, \qquad m_2 = a - ib$$

The solution is then

$$\begin{aligned} C &= K_1e^{m_1x} + K_2e^{m_2x} \\ &= K_1e^{(a+ib)x} + K_2e^{(a-ib)x} \\ &= K_1e^{ax}e^{ibx} + K_2e^{ax}e^{-ibx} \\ &= e^{ax}(K_1e^{ibx} + K_2e^{-ibx}) \end{aligned} \tag{5.109}$$

To simplify this further we introduce Euler's formulas

$$e^{ibx} = \cos bx + i \sin bx$$

$$e^{-ibx} = \cos bx - i \sin bx$$

Substituting these relations into the solution for C yields

$$C = e^{ax}[(K_1 + K_2) \cos bx + i(K_1 - K_2) \sin bx] \tag{5.110}$$

Since the terms $K_1 + K_2$, and $i(K_1 - K_2)$ are also constants, we write the solution in the form

$$C = e^{ax}(K_1 \cos bx + K_2 \sin bx) \tag{5.111}$$

The results for the determination of the complementary solution with the three cases (A), (B), and (C) are determined as follows: To find the solution to the differential equation

$$\frac{d^2C}{dx^2} + P\frac{dC}{dx} + QC = 0$$

determine the roots of the quadratic equation

$$(m^2 + Pm + Q) = 0$$

These roots are

$$m_1 = -\frac{P}{2} + \frac{1}{2}\sqrt{(P^2 - 4Q)}$$

$$m_2 = -\frac{P}{2} - \frac{1}{2}\sqrt{(P^2 - 4Q)}$$

If these roots are:

(A) real and unequal $(P^2 - 4Q) > 0$, the solution is

$$C = K_1 e^{m_1 x} + K_2 e^{m_2 x}$$

(B) real and equal $(P^2 - 4Q) = 0$, the solution is

$$C = (K_1 x + K_2)e^{mx}$$

where $m = m_1 = m_2 = -P/2$, and

(C) complex and unequal $(P^2 - 4Q) < 0$, the solution is

$$C = e^{ax}(K_1 \cos bx + K_2 \sin bx)$$

where $a = -P/2$, and $b = \sqrt{(4Q - P^2)}/2$.

Example 5.33. Solve the differential equation

$$\frac{d^2C}{dx^2} - \frac{dC}{dx} - 2C = 0$$

The auxiliary equation is

$$(m^2 - m - 2) = 0$$

with roots $m_1 = -1$ and $m_2 = 2$. The roots are real and unequal; therefore the solution is

$$C = K_1 e^{-x} + K_2 e^{2x}$$

Example 5.34. Solve the differential equation

$$\frac{d^2C}{dx^2} + 2\frac{dC}{dx} - C = 0$$

The auxiliary equation is

$$(m^2 + 2m + 1) = 0$$

with the real, equal roots $m_1 = m_2 = -1$. The solution is

$$C = (K_1x + K_2)e^{-x}$$

Example 5.35. Solve the differential equation

$$\frac{d^2C}{dx^2} + 2\frac{dC}{dx} + 2C = 0$$

with roots $m_1 = -1 + i$ and $m_2 = -1 - i$. The roots are complex and unequal, so we find that $a = -1$ and $b = 1$. Thus the solution is

$$C = e^{-x}(K_1 \cos x + K_2 \sin x)$$

Complementary Higher Order Equation with Constant Coefficients. For complementary equations of higher order, we try the same technique as used for the second-order equations.

Example 5.36. Solve the differential equation

$$\frac{d^4C}{dx^4} + 3\frac{d^3C}{dx^3} - 3\frac{d^2C}{dx^2} - 7\frac{dC}{dx} + 6 = 0$$

The auxiliary equation, obtained by substituting $C = e^{mx}$ in the equation, is

$$(m^4 + 3m^3 - 3m^2 - 7m + 6) = 0$$

and has the roots $m_1 = m_2 = 1$, $m_3 = -2$, and $m_4 = -3$. We treat the real double roots as before, and conclude that $(K_1x + K_2)e^x$ is a solution. The other two real roots are equal, and therefore we conclude that the other two solutions are K_3e^{-2x} and K_4e^{-3x}. Therefore, the complementary solution containing the expected four constants is

$$C = (K_1x + K_2)e^x + K_3e^{-2x} + K_4e^{-3x}$$

This solution can be verified by substitution back into the differential equation. Since the equation is linear, we can apply the principle of superposition, valid for all linear differential equations, and independently verify each of the four solutions xe^x, e^x, e^{-2x}, and e^{-3x}.

Thus we have seen that the solution technique which was valid for the general second-order differential equation was also valid for the fourth-order equation in our example. It can be shown that this technique is also valid for the general nth-order linear differential equation with constant coefficients.

To solve the general nth-order differential equation, we substitute $y = e^{mx}$ into the equation to obtain the auxiliary equation for m. Actually, the reader should have observed by now that this substitution step for the auxiliary equation may be bypassed, and all that is necessary is to replace each derivative of the equation with m to a power that is equal to the order of the derivative. The complementary nth order differential equation with constant coefficients is

$$\frac{d^nC}{dx^2} + P_1\frac{d^{n-1}C}{dx^{n-1}} + \ldots + P_{n-1}\frac{dC}{dx} + Pn = 0 \tag{5.112}$$

The auxiliary equation is

$$m^n + P_1m^{n-1} + \ldots + P_{n-1}m + P_n = 0 \tag{5.113}$$

This equation will in general have a combination of roots that are real and unequal, real and equal, complex and unequal, and complex and equal. Therefore, to illustrate the treatment of the higher order equation we need only treat various types of roots. In the following examples, we give the roots of the auxiliary equation obtained from an indicated differential equation:

Example 5.37. Given a third-order differential equation whose roots are

$$m_1 = 1,$$

$$m_2 = ½\,(-1 + i\sqrt{3})$$

and

$$m_3 = ½\,(-1 - i\sqrt{3}).$$

Recalling that the complex roots can be written as $m_2 = a + ib$ and $m_3 = a - ib$, where $a = -1/2$ and $b = \sqrt{3}$, the solution is

$$C = K_1e^x + e^{-x/2}K_3(\cos\sqrt{3}\,x + K_4\sin\sqrt{3}\,x)$$

Example 5.38. Given an eighth-order differential equation whose roots are $m_1 = m_2 = m_3 = 2$, $m_4 = m_5 = 0$, $m_6 = 3$, $m_7 = 2 + 3i$, and $m_8 = 2 - 3i$: the triple root two leads to the solution $(K_1x^2 + K_2x + K_3)e^{2x}$; the double root zero leads to the solution $(K_4x + K_5)e^{0x} = K_4x + K_5$; the other roots yield solutions similar to example 5.37. The complementary solution is then

$$C = (K_1x^2 + K_2x + K_3)e^{2x} + K_4x + K_5 + K_6e^{3x}$$
$$+ e^{2x}(K_7\cos 3x + K_8\sin 3x)$$

Example 5.39. Given a fourth-order differential equation whose roots are $m_1 = m_2 = m_3 = m_4 = 2 + 3i$: Treating these repeated roots in the same manner as repeated real roots, we write the complementary solution as

$$C = (K_1x + K_2)e^{2x}(K_3 \cos 3x + K_4 \sin 3x)$$

For the case of an nth-order differential equation with n repeated roots—for example, $m_1 = m_2 = \ldots = m_{n-1} = m_n = m$—we multiply the one exponential solution e^{mx} by a polynomial of order $n - 1$ containing n arbitrary constants. Thus for n repeated roots we have the solution

$$C = (K_1x^{n-1} + K_2x^{n-2} + \ldots + K_{n-1}x + K_n)e^{mx} \qquad (5.114)$$

For the case of n repeated complex roots we multiply the one complex solution which contains two arbitrary constants by a similar polynomial.

The Particular Solution. Method of Undetermined Coefficients. We have discussed how to obtain the complementary solution, which was a general solution to the complementary equation. We now seek a particular solution to the complete differential equation with constant coefficients

$$\frac{d^nC}{dx^2} + P_1\frac{d^{n-1}C}{dx^{n-1}} + \ldots + P_{n-1}\frac{dC}{dx} + P_nC = R(x) \qquad (5.115)$$

Recall that a particular solution to equation (5.115) is *any solution* regardless of how special, and contains no arbitrary constants. The general solution to complete equations is the sum of the complementary solution and a particular solution. We proceed with our discussion by considering the second-order differential equation:

$$\frac{d^2C}{dx^2} + P\frac{dC}{dx} + Q = R(x) \qquad (5.116)$$

The method of undetermined coefficients for solving equations can be considered sophisticated guessing. We simply assume particular solutions according to the form of $R(x)$. The method is limited to certain forms of $R(x)$, however, such as combinations of polynomials, exponentials, sines, and cosines. The following examples will illustrate the method:

Example 5.40. Solve the differential equation

$$\frac{d^2C}{dx^2} + 5\frac{dC}{dx} - 6C = 6e^{3x}$$

From our previous treatment of exponentials, recall that e^{3x} is not changed by differentiation. All derivatives merely reproduce the function times a constant. Therefore, we assume a function of the form $C_P = ae^{3x}$. Substituting this form of C into the differential equation gives

$$9ae^{3x} + 15ae^{3x} - 6ae^{3x} = 6e^{3x}$$

or

$$18ae^{3x} = 6e^{3x}$$

therefore $a = 1/3$. Thus a particular solution is

$$C_p = \frac{1}{3}e^{3x}$$

which can be verified by substitution. We had previously determined the complementary solution as

$$C_c = K_1e^{-2x} + K_2e^{-3x}$$

The general solution to the complete equation in the example therefore is

$$C = C_p + C_c = \frac{1}{3}e^{3x} + K_1e^{-2x} + K_2e^{-3x}$$

Example 5.41. Solve the differential equation

$$\frac{d^2C}{dx^2} - \frac{dC}{dx} - 2C = 2x^2 + 3x + 1$$

What forms of C, when differentiated, will yield a polynomial? The answer, of course, is another polynomial. Polynomials contain terms such as x^n which upon each differentiation have their exponents reduced by one, yielding nx^{n-1}. Since our equation contains the term C, the highest order polynomial term that we could assume is x^2. We, therefore, assume the general second-degree polynomial $C_p = ax^2 + bx + d$ as our particular solution. Substituting in the differential equation and rearranging terms yields

$$-2ax^2 - (2a + 2b)x + (2a - b - 2d) = 2x^2 + 3x + 1$$

Thus, equating coefficients of like powers of x,

$$-2a = 2, \qquad -(2a + 2b) = 3, \qquad (2a - b - 2d) = 1$$

Solving, we get

$$a = -1, \qquad b = -\frac{1}{2}, \qquad d = -\frac{5}{4}$$

and our particular solution is

$$C_p = -x^2 - \frac{1}{2}x - \frac{5}{4}$$

Example 5.42. Solve the differential equation

$$\frac{d^2C}{dx^2} - \frac{dC}{dx} - 2C = 2\cos 2x$$

The determination of particular solutions for differential equations containing nonhomogeneous trigonometric terms such as $\sin x$ and $\cos x$ is similar to that for equations containing exponential terms. The functions $\sin x$ and $\cos x$ reproduce the negative of themselves after two differentiations, whereas the exponential function reproduces itself after each differentiation. Thus for the exponential, the order of the derivatives in the equation was not of particular concern. However, odd-ordered derivatives of $\cos x$ and $\sin x$ will produce $\sin x$ and $\cos x$, respectively. Thus the assumption $C_p = a \cos 2x$ as a particular solution for the above differential equation will yield $-2a \sin 2x$ on the left-hand side of the equation. Since there is no like function on the right-hand side, this term cannot be eliminated from the equation. We therefore choose as a particular solution

$$C_p = a \cos 2x + b \sin 2x$$

which upon substitution into the equation and rearrangment of terms yields

$$(-6a - 2b) \cos 2x + (2a - 6b) \sin 2x = 2 \cos 2x$$

Equating coefficients of like functions,

$$(-6a - 2b) = 2, \qquad (2a - 6b) = 0$$

Solving, we get

$$a = -\frac{3}{10}, \qquad b = -\frac{1}{10}$$

and the particular solution is

$$C_p = -\frac{3}{10} \cos 2x - \frac{1}{10} \sin 2x$$

Summarizing what we have seen in our examples, we can state that when a differential equation is in the form of equation (5.116), and $R(x)$ contains terms such as e^{mx}, x^n ($n = 0, 1, 2, \ldots$), $\sin kx$, $\cos kx$, we assume as a particular solution corresponding terms ae^{mx}, bx^n, $d \sin kx$, and $f \cos kx$, where a, b, d, and f are constants. For example, if the nonhomogeneous term is of the form

$$R(x) = 3 \sin 2x + 2e^{x/2} + 2x^3 + 6x + 4$$

we would assume a particular solution containing $a \sin 2x + b \cos 2x$ to treat the term $3 \sin 2x$, $de^{x/2}$ to treat the term $2e^{x/2}$, and $fx^3 + gx^2 + hx + 1$ to treat the term $2x^3 + 6x + 4$—that is, the appropriate form of the particular solution would be

$$C_p = a \sin 2x + b \cos 2x + de^{x/2} + fx^3 + gx^2 + hx + 1$$

To determine the coefficients in this expression, we would substitute C_p back into the differential equation and equate coefficients of like functions as in our examples.

Compound Terms. It is possible to solve for the particular solution when $R(x)$ contains compound terms such as $x^n e^{mx} \cos kx$ and $x^n e^{mx} \sin kx$. The assumed forms should be $(a_0 x^n + a_1 x^{n-1} + \ldots + a_{n-1}x + a_n)e^{mx} \cos kx$ and $(b_0 x^n + b_1 x^{n-1} + \ldots + b_{n-1}x + b_n)e^{mx} \sin kx$, respectively.

Example 5.43. Solve the differential equation

$$\frac{d^2C}{dx^2} - \frac{dC}{dx} - 2C = 2x^2 e^x$$

We assume a particular solution of the form

$$C_p = (ax^2 + bx + d)e^x$$

and substitute this into the equation to obtain

$$-2ax^2 e^x + (2a - 2b)xe^x + (2a + b - 2d)e^x = 2x^2 e^x$$

Equating coefficients of like terms and solving, we get

$$a = -1, \qquad b = -1, \qquad d = -\frac{3}{2}$$

The particular solution is therefore

$$C_p = -x^2 e^x - xe^x - \frac{3}{2}e^x$$

Even for a relatively simple compound term such as $2x^2 e^x$ the solution procedure is tedious. For a nonhomogeneous term such as

$$R(x) = 3x^3 e^{2x} \sin 2x$$

The assumed solution form should be

$$C_p = (a_0 x^3 + a_1 x^2 + a_2 x + a_3)e^{2x} \sin 2x$$
$$+ (b_0 x^3 + b_1 x^2 + b_2 x + b_3)e^{2x} \cos 2x \qquad (5.117)$$

The reader who has been working through the examples so far should be able to see how lengthy the manipulation would be in this case. To avoid the tedious calculations that would be required to determine the particular solution, alternate solution methods should be considered.

Example 5.44. Solve the differential equation

$$\frac{d^2C}{dx^2} - \frac{dC}{dx} - 2C = e^x \cos 2x$$

We assume the form

$$C_p = e^x(a \cos 2x + b \sin 2x)$$

which upon substitution into the equation yields

$$(-6a + 2b)e^x \cos 2x + (-6b - 2a)e^x \sin 2x = e^x \cos 2x$$

Equating coefficients of like terms and solving, we get

$$a = -\frac{3}{20}, \qquad b = \frac{1}{20}$$

The particular solution is thus

$$C_p = \frac{1}{20}e^x(-3 \cos 2x + \sin 2x)$$

Special Cases. There is a breakdown in the general procedure we have been following when $R(x)$ contains a term which is a solution of the complementary equation.

Example 5.45. Solve the differential equation

$$\frac{d^2C}{dx^2} - \frac{dC}{dx} - 2C = e^{-x}$$

Following the usual procedure, we would try $C_p = ae^{-x}$, substitute this into the equation, and equate coefficients of like terms to solve for a. However, the substitution would yield

$$+ae^{-x} + ae^{-x} - 2ae^{-x} = e^{-x}$$

or

$$0 = e^{-x}$$

which is nonsense.

If we recall that the complementary solution of this equation was

$$C_c = K_1e^{-x} + K_2e^{2x}$$

we see that the nonhomogeneous term $R(x) = e^{-x}$ duplicates a term in the complementary solution. This example typifies the difficulty that occurs in cases such as this. Guided by a somewhat similar problem that occurred with complementary equations whose auxiliary equations contain multiple roots, we assume the form

$$C_p = axe^{-x}$$

Substituting back into the differential equation yields

$$-3ae^{-x} = e^{-x}$$

Thus $a = -1/3$ and our particular solution is

$$C_p = -\frac{1}{3}xe^{-x}$$

For the general case where $R(x)$ duplicates a term in the complementary solution, we multiply the expected particular C_p by the "lowest" power of x, which produces a term not already appearing in the complementary solution.

Example 5.46. Solve the differential equation

$$\frac{d^2C}{dx^2} + 4C = \cos 2x$$

The complementary solution can be shown to be

$$C_c = K_1 \cos 2x + k_2 \sin 2x$$

Thus by our previous discussion, it would be useless to assume a particular solution of the form $C_p = a \cos 2x + b \sin 2x$, since it would vanish upon substitution into the differential equation. We therefore assume that

$$C_p = ax \cos 2x + bx \sin 2x$$

Substituting C_p into the equation yields

$$4b \cos 2x - 4a \sin 2x = \cos 2x$$

Thus

$$b = \frac{1}{4}, \qquad a = 0$$

and the particular solution is

$$C_p = \frac{1}{4} x \sin 2x$$

We summarize the method of undetermined coefficients here. For the differential equation

$$\frac{d^2C}{dx^2} + P\frac{dC}{dx} + Q = R(x)$$

where P and Q are constants, a particular solution C_p can be found when $R(x)$ equals:

Ax^n, by assuming $C_p = a_0x^n + a_1x^{n-1} + \ldots + a_{n-1}x + a_n$

Ae^{mx}, by assuming $C_p = ae^{mx}$

$A \cos kx$ or $A \sin kx$, by assuming $C_p = a \cos kx + b \sin kx$

$Ax^ne^{mx} \cos kx$ or $Ax^ne^{mx} \sin kx$, by assuming

$$C_p = (a_0x^n + a_1x^{n-1} + \ldots + a_{n-1}x + a_n)x^ne^{mx} \cos kx + (b_0x^n + b_1x^{n-1} + \ldots + b_{n-1}x + b_n)x^ne^{mx} \sin kx$$

where A, n, and m are known constants ($n = 0, 1, 2, 3, \ldots.$), and a, b, a_0, $a_1, \ldots, a_n$, and $b_0, b_1, \ldots, b_n$ are constant coefficients.

If any of these assumed terms duplicates a term appearing in the complementary solution, multiply these terms by the *least* power of x, which will convert these terms to ones that are not duplicated in the complementary solution. Substitute the assumed form for C_p into the differential equation and equate coefficients of like terms to evaluate the coefficients.

Although we have used the second-order linear differential equation throughout our discussion of the method of undetermined coefficients, equations of higher order can also be treated in the same manner. We conclude this section with an example of the method applied to a fourth-order differential equation.

Example 5.47. Solve the differential equation

$$\frac{d^4C}{dx^4} + 3\frac{d^3C}{dx^3} - 3\frac{d^2C}{dx^2} - 7\frac{dC}{dx} + 6 = e^{2x}$$

We assume a solution of the form $C_p = ae^{2x}$. Substituting this form of C into the equation gives

$$ae^{2x}[16 + 3(8) - 3(4) - 7(2) + 6] = e^{2x}$$

or

$$20a = 1$$

and therefore $a = 1/20$. Thus a particular solution is

$$C_p = \frac{1}{20}e^{2x}$$

The Method of Variation of Parameters. The method of undetermined coefficients was limited to differential equations whose right-hand side was of a special type. Also, in the case of certain differential equations the method was tedious, and it was mentioned that an alternate solution method would be considered. The method of *variation of parameters* is effective for all linear differential equations, even those equations whose coefficients are not constant, whenever we can solve the complementary equation. The use of this method requires a knowledge of integration, whereas the method of undetermined coefficients utilized differentiation. Thus the term *particular integral* is especially appropriate for the solution obtained by variation of parameters.

Consider the general second-order differential equation given by

$$\frac{d^2C}{dx^2} + P(x)\frac{dC}{dx} + Q(x)C = R(x) \tag{5.118}$$

The complementary solution of the corresponding reduced equation

$$\frac{d^2C}{dx^2} + P(x)\frac{dC}{dx} + Q(x)C = 0 \tag{5.119}$$

can be written as

$$C = K_1C_1(x) + K_2C_2(x) \tag{5.120}$$

where K_1 and K_2 are arbitrary constants and $C_1(x)$ and $C_2(x)$ are known independent solutions of the complementary equation. The basis of the method of variation of parameters is to replace the known constants K_1 and K_2 with two unknown functions, $K_1(x)$ and $K_2(x)$, that vary with x and require that the new expression

$$C_p = K_1(x)\,C_1(x) + K_2(x)\,C_2(x) \tag{5.121}$$

be a solution of equation (5.118). The method thus assumes that the arbitrary constants K_1 and K_2 vary and are therefore called parameters—hence the name variation of parameters.

We have two unknowns, K_1 and K_2, to determine, but only the one equation (5.118), and we are therefore free to choose another equation, or condition on K_1 and K_2, at our convenience. We form the first derivative of equation (5.121) as

$$C_p' = K_1'(x)\,C_1(x) + K_1(x)\,C_1'(x) + K_2'(x)\,C_2(x) + K_2(x)\,C_2'(x) \tag{5.122}$$

To reduce the complexity of satisfying equation (5.118), we take as our second condition one which eliminates the derivatives of the unknowns in equation (5.122). Thus we set

$$K_1'(x)\,C_1(x) + K_2'(x)\,C_2(x) = 0 \tag{5.123}$$

and equation (5.122) becomes

$$C_p' = K_1(x)\, C_1'(x) + K_2(x)\, C_2'(x) \tag{5.124}$$

Differentiating again,

$$C_p'' = K_1'(x)\, C_1'(x) + K_1(x)\, C_1''(x) + K_2'(x)\, C_2'(x) + K_2(x)\, C_2''(x) \tag{5.125}$$

Substituting C, C', and C'' into equation (5.118), we obtain after simplification

$$\begin{aligned} &K_1(x)[C_1''(x) + P(x)\, C_1'(x) + Q(x)\, C_1(x)] + K_2(x)[C_2''(x) \\ &\quad + P(x)\, C_2'(x) + Q(x)\, C_2(x)] + K_1'(x)\, C_1'(x) + K_2'(x)\, C_2'(x) = R(x) \end{aligned} \tag{5.126}$$

Since C_1 and C_2 are solutions to the complementary equation, the quantities in brackets vanish, and we have the second equation for K_1 and K_2:

$$K_1'(x)\, C_1'(x) + K_2'(x)\, C_2'(x) = R(x) \tag{5.127}$$

Solving equations (5.123) and (5.127) for K_1' and K_2' yields

$$K_1'(x) = -\frac{C_2(x)}{W(x)}\, R(x), \qquad K_2'(x) = \frac{C_1(x)}{W(x)}\, R(x) \tag{5.128}$$

where

$$W(x) = C_1(x)\, C_2'(x) - C_1'(x)\, C_2(x)$$

is called the *wronskian* of the solutions C_1 and C_2. It is known that if the two solutions C_1 and C_2 are independent, the wronskian cannot vanish, and we therefore expect no difficulty with our solution thus far. Integrating, we get

$$K_1(x) = -\int \frac{C_2(x)R(x)}{W(x)}\, dx, \qquad K_2(x) = \int \frac{C_1(x)\, R(x)}{W(x)}\, dx \tag{5.129}$$

and the particular solution is given by

$$C_p = -C_1(x) \int \frac{C_2(x)\, R(x)}{W(x)}\, dx + C_2(x) \int \frac{C_1(x)\, R(x)}{W(x)}\, dx \tag{5.130}$$

Example 5.48. Solve the differential equation

$$\frac{d^2C}{dx^2} - \frac{dC}{dx} - 2C = e^{-x}$$

The complementary solution of this equation is known to be

$$C_c = K_1 C_1(x) + K_2 C_2(x)$$

where

$$C_1(x) = e^{-x}, \qquad C_2(x) = e^{2x}$$

Now the wronskian can be evaluated as

$$W(x) = e^{-x}(2e^{2x}) - (-e^{-x})e^{2x}$$
$$= 3e^{-x}e^{2x}$$

and therefore

$$K_1(x) = -\int \frac{e^{2x}e^{-x}}{3e^{-x}e^{2x}}\,dx = -\frac{1}{3}\int dx = -\frac{x}{3}$$

and

$$K_2(x) = \int \frac{e^{-x}e^{-x}}{3e^{-x}e^{2x}}\,dx = \frac{1}{3}\int e^{-3x}\,dx = -\frac{e^{-3x}}{9}$$

Thus the particular integral is

$$C_p = e^{-x}\left(-\frac{x}{3}\right) + e^{2x}\left(-\frac{e^{-3x}}{9}\right)$$
$$= -\frac{x}{3}e^{-x} - \frac{1}{9}e^{-x}$$

The general solution to the complete equation is

$$C = K_1e^{-x} + K_2e^{2x} - \frac{x}{3}e^{-x} - \frac{1}{9}e^{-x}$$

Since they both are constants multiplying e^{-x}, the last term in this expression can be combined with the first term to yield

$$C = K_1e^{-x} + K_2e^{2x} - \frac{x}{3}e^{-x}$$

The term $(-1/9)e^{-x}$ would be a solution to the reduced equation and thus adds nothing to the particular solution. Note that in this differential equation, $R(x)$ duplicated the term e^{-x} in the complementary solution.

Example 5.49. Solve the differential equation

$$\frac{d^2C}{dx^2} + 4C = \cos 2x$$

which has the complementary solution

$$C_c = K_1C_1(x) + K_2C_2(x)$$

where

$$C_1(x) = \cos 2x, \qquad C_2(x) = \sin 2x$$

The wronskian is

$$W(x) = \cos 2x\,(2\cos 2x) - (-2\sin 2x)(\sin 2x)$$
$$= 2$$

and therefore

$$K_1(x) = -\frac{1}{2}\int \sin 2x \cos 2x\, dx$$
$$= -\frac{1}{8}\sin^2 2x$$

and

$$K_2(x) = \frac{1}{2}\int \cos 2x \cos 2x\, dx$$
$$= \frac{1}{4}\left(x + \frac{1}{2}\sin 2x \cos 2x\right)$$

Thus the particular integral is

$$C_p = \cos 2x\left(-\frac{1}{8}\sin^2 2x\right) + \sin 2x\left(\frac{x}{4} + \frac{1}{8}\sin 2x \cos 2x\right)$$
$$= \frac{1}{4}x \sin 2x$$

The general solution—the sum of the complementary and particular solutions—is therefore

$$C - K_1 \cos 2x + K_2 \sin 2x + \frac{1}{4}x \sin 2x$$

Example 5.50. Solve the differential equation

$$\frac{d^2C}{dx^2} - \frac{dC}{dx} - 2C = xe^{-x}$$

This is the same equation as in example 5.48, and thus we have

$$C_1(x) = e^{-x}, \qquad C_2(x) = e^{2x}, \qquad W(x) = 3e^{-x}e^{2x}$$

therefore

$$K_1(x) = -\int \frac{e^{2x}(x\,e^{-x})}{3e^{-x}e^{2x}}\,dx = -\frac{1}{3}\int x\,dx = -\frac{1}{6}x^2$$

and

$$K_2(x) = \int \frac{e^{-x}(xe^{-x})}{3e^{-x}e^{2x}}\,dx = \frac{1}{3}\int xe^{-3x}\,dx = -\frac{e^{-3x}}{27}(3x + 1)$$

The particular integral can now be written as

$$C_p = e^{-x}\left(-\frac{1}{6}x^2\right) + e^{2x}\left[-\frac{e^{-3x}}{27}(3x + 1)\right]$$

or simplifying,

$$C_p = -e^{-x}\left(\frac{1}{6}x^2 + \frac{1}{9}x + \frac{1}{27}\right)$$

Example 5.51. Solve the differential equation

$$\frac{d^2C}{dx^2} - 4C = \ln x$$

The complementary solution is

$$C_c = K_1C_1(x) + K_2C_2(x)$$

where

$$C_1(x) = e^{2x} \qquad C_2(x) = e^{-2x}$$

The wronskian is

$$W(x) = e^{2x}\,(-2e^{-2x}) - 2e^{2x}e^{-2x}$$
$$= 4$$

therefore

$$K_1(x) = -\frac{1}{4}\int e^{-2x}\ln x\,dx = \frac{1}{8}\,e^{-2x}\ln x + \frac{1}{2}\int \frac{e^{-2x}}{x}dx$$

and

$$K_2(x) = \frac{1}{4}\int e^{2x}\ln x\,dx = \frac{1}{8}e^{2x}\ln x - \frac{1}{2}\int \frac{e^{2x}}{x}dx$$

Thus the particular solution is

$$C_p = e^{2x}\left(\frac{1}{8}e^{-2x}\ln x + \frac{1}{2}\int \frac{e^{-2x}}{x}\,dx\right)$$
$$+ e^{-2x}\left(\frac{1}{8}e^{2x}\ln x - \frac{1}{2}\int \frac{e^{2x}}{x}\,dx\right)$$

$$= \frac{1}{4} \ln x + \frac{1}{2} e^{2x} \left(\int \frac{e^{-2x}}{x} dx \right) - \frac{1}{2} e^{-2x} \left(\int \frac{e^{2x}}{x} dx \right)$$

This example illustrates a case where the resulting integrals cannot be evaluated in terms of simple functions—that is,

$$\int \frac{e^{mx}}{x} dx = \ln x + mx + \frac{m^2x^2}{2(2!)} + \frac{m^3x^3}{3(3!)} + \ldots .$$

The method of variation of parameters is applicable to a wide range of linear differential equations and may be extended to higher order equations.

Remarks

We have introduced two methods yielding particular solutions that can be used in conjunction with the methods for determining the complementary solutions to obtain the general solution to the complete equation. These methods should be sufficient to treat linear differential equations with constant coefficients. Certain special equations that had nonconstant coefficients, and others that were nonlinear were also treated.

There are many other methods that are used to solve differential equations, but we have selected those which we feel would be most applicable to the reader of this text. Substitutions and transformations have been used to reduce certain complicated differential equations to ones that could be treated by previously explained methods. Again, we have been selective and had to omit many other clever transformations that could be utilized to solve higher order equations. There is, however, one special differential equation with variable coefficients that can be reduced, by a change of variables, to a linear equation with constant coefficients. This equation is called the *Euler* or *Cauchy* differential equation. A second-order Euler differential equation can be written as

$$x^2 \frac{d^2C}{dx^2} + P_1 x \frac{dC}{dx} + P_2 C = R(x) \tag{5.131}$$

where P_1 and P_2 are constants. Note that the nth-order derivative is multiplied by the nth power of x. Similarly, for the nth-order differential equation when $n > 2$, we would have terms such as $x^n \, d^nC/dx^n$.

The change of variable $x = e^t$ or $t = \ln x$ reduces the Euler equation to an equation in C and t with constant coefficients.

Example 5.52. Solve the differential equation

$$x^2 \frac{d^2C}{dx} - 2C = 6$$

Now

$$\frac{dC}{dx} = \frac{dC}{dt} \times \frac{dt}{dx} = \frac{1}{x} \times \frac{dC}{dt}$$

and therefore

$$\begin{aligned}\frac{d^2C}{dx^2} &= \frac{d}{dx}\left(\frac{1}{x} \times \frac{dC}{dt}\right) \\ &= \frac{1}{x} \times \frac{d}{dx}\left(\frac{dC}{dt}\right) - \frac{1}{x^2} \times \frac{dC}{dt} \\ &= \frac{1}{x} \times \frac{d}{dt}\left(\frac{dC}{dt}\right)\frac{dt}{dx} - \frac{1}{x^2} \times \frac{dC}{dt} \\ &= \frac{1}{x^2}\left(\frac{d^2C}{dt^2} - \frac{dC}{bt}\right)\end{aligned}$$

thus

$$x^2 \frac{d^2C}{dx^2} = \frac{d^2C}{dt^2} - \frac{dC}{dt}$$

The original differential equation can now be written as

$$\frac{d^2C}{dt^2} - \frac{dC}{dt} - 2C = 6$$

which is a linear differential equation with constant coefficients and can be readily solved by methods previously introduced.

Partial Differential Equations

Partial differential equations have previously been defined as equations that contain two or more independent variables. The derivatives in these equations are called *partial derivatives*. Identifying and classifying the partial differential equation according to order, degree, linearity, and homogeneity is the same as for ordinary differential equations. Unfortunately, the presence of more independent variables makes the solution of partial differential equations more complicated, so our coverage, in keeping with the general format of this text, is brief.

We have seen how partial differential equations arise from physical problems in chapter 4. Recall, for example, the BOD model in the form

$$\frac{\partial L}{\partial t} + P_1 \frac{\partial L}{\partial x} + P_2 L = L_a \tag{5.132}$$

where P_1, P_2, and L_a are constants that depend on the physical system, and L is a function of the two independent variables x and t. Recall also the microbial die-off model

$$\frac{\partial C}{\partial t} + P_1 \frac{\partial C}{\partial x} + P_2 \frac{\partial C}{\partial y} + P_3 \frac{\partial C}{\partial z} + P_4 C = 0 \tag{5.133}$$

where P_1, P_2, P_3, and C may all be functions of the four independent variables x, y, z, and t, and P_4 is a function of z and contains constants dependent on the physical system. These equations are difficult to solve and have been treated in previous chapters by considering special cases in which the partial differential equation reduced to an ordinary differential equation.

For the purpose of discussion, consider the simple partial differential equation

$$\frac{\partial C(x,t)}{\partial x} = f(x,t) \tag{5.134}$$

which is analagous to the simplest differential equation treated in chapter 5:

$$\frac{dC}{dx} = f(x) \tag{5.135}$$

Since we have the derivative of C with t held constant in equation (5.134), we determine C by integrating with t held constant. Thus

$$C = \int f(x,t)\,\partial x + K_1 \tag{5.136}$$

where the symbol ∂x indicates that the integration is with respect to x, and K_1 is an arbitrary constant of integration with respect to x, but may be a function of t—that is, $K_1 = K_1(t)$. We write

$$C = \int f(x,t)\,\partial x + K_1(t) \tag{5.137}$$

Thus our solution of equation (5.134) contains an *arbitrary function*, the first apparent difficulty that arises due to the presence of more than one independent variable.

Example 5.53. Solve the partial differential equation

$$\frac{\partial C(x,t)}{\partial x} = xe^{-t}$$

Integration with respect to x only yields

$$C(x,t) = \frac{x^2 e^{-t}}{2} + K_1(t)$$

To solve for $K_1(t)$ we would need a condition on $C(x,t)$. For example, assume we had the condition that

$$C(1,t) = 2e^{-t}$$

Substituting $x = 1$ into our expression for $C(x,t)$ and using our condition yields

$$\frac{e^t}{2} + K_1(t) = 2e^{-t}$$

and therefore

$$K_1(t) = \frac{3}{2}e^{-t}$$

Note that the boundary condition on C was in terms of a function of t, whereas for an ordinary differential equation a boundary condition was a specific numerical value.

If C had been a function of three variables—for example, $C = C(x,y,t)$—then the arbitrary function of integration would have the form $K_1(y,t)$. This could be similarly extended to the case of more than three independent variables.

Since we have limited our discussion of partial differential equations to just the brief description here, the analyst will have to refer to the many excellent textbooks on this subject if he encounters a partial differential equation other than the most simple type. If, however, the physical system dictates that special cases of the equation reduce it to an ordinary differential equation, then the techniques discussed in this chapter should be applicable.

6 Elements of Statistics

Earl Rich

Introduction

Our understanding of nature and the consequences of our manipulation of nature is based on the ability to detect permanencies—the phenomena which we tend to observe with some degree of regularity and predictability. Whenever a permanency is perceived we, as scientists, try to erect some sort of intellectual framework which explains the permanency and allows for the use of our new-found knowledge. In other words, we detect a phenomenon, make a model, and use the model as a basis of inductive behavior. The usefulness of the model will be limited by the degree to which the model is an accurate reflection of nature and by the quality of the information used in the model. As a guide to this intellectual effort some sort of objective measure is needed that will tell us how good the information is and how well our model fits reality.

For the most part scientific application of statistical techniques fall into two categories: the estimation of parametric values, and the testing of hypotheses. To develop a few simple and practical methods we need certain definitions and some theoretical background. It would not be feasible to attempt a comprehensive coverage here. Statistical theory is a massive and actively growing discipline, and the armament of methodology is already massive with many powerful techniques and many highly specialized procedures. We primarily concern ourselves in this chapter with data evaluation and comparison of predictions of theoretical models with observations of the real world.

Basic Concepts

The most common problem encountered with data evaluation is a consequence of the fact that we find ourselves trying to estimate parametric values that cannot be measured without some uncontrollable random component. For example, we may observe or measure a variable x that is based on a true but unknown parametric value μ and a random component ϵ.

$$x_i = \mu + \epsilon_i \tag{6.1}$$

However we try, we cannot eliminate the random component (ϵ). It may be minimized by ingenious instrumentation, but because this random component is often itself rather complicated and difficult to control, we often do better to obtain the best possible evaluation of ϵ. Assumptions made about ϵ and its distribution are of particular theoretical importance.

Considering that x has a random component, it can be spoken of as a *random variable*. It will take on its various possible values with some probability, and this distribution is a *probability distribution*. We deal with only a few of the many sorts of distributions, but those selected are useful and powerful and require relatively few simplifying assumptions.

Descriptive Statistics

The process of collecting quantitative information and then summarizing and presenting it is the area of descriptive statistics. The objective is to describe in such a way that unambiguous communication is possible and to provide a basis for the ultimate aim of making inferences about the data. A measure obtained from a set of observations or data is termed a *statistic*. The set of observations is a *sample*, and it is a subset of some totality of possible observations called a *population* or *universe*.

By way of illustration, we examine a number of problems of observation one could make about a river:

(A) If we were interested in the rate of flow of a river, we might use some measuring device that would tell the velocity at which water passed a certain point in space. A simple observation at a sample point would not provide sufficient information to reach any conclusions about the volume of flow in the river. The main problem here is that the flow varies from point to point in any given cross section of a stream. Figure 6–1 shows a hypothetical cross section of a stream. The actual conditions in a stream could be quite complicated. Here the lines connect points of equal velocity of flow. The greatest velocity is in the region of A, the lowest is at B and C because of bottom friction, while the velocity at D would have an intermediate value. If we wanted the observational data to calculate the value of flow past a point, we would want to sample from all areas of the river and obviously would want the observations to be taken from the various flow-rate zones in a manner such that the number from each zone was proportional to the cross-sectional area of the zone. Here we might want a large number of samples taken at randomly selected points on the cross-sectional plane in such a way that every point on the plane had an equal chance of being included as a sample point. Clearly, the problem here is much too complicated to solve by just going out in the middle of the river and dropping in a flow meter. Also, we haven't even mentioned season-to-season, or year-to-year variability of flow.

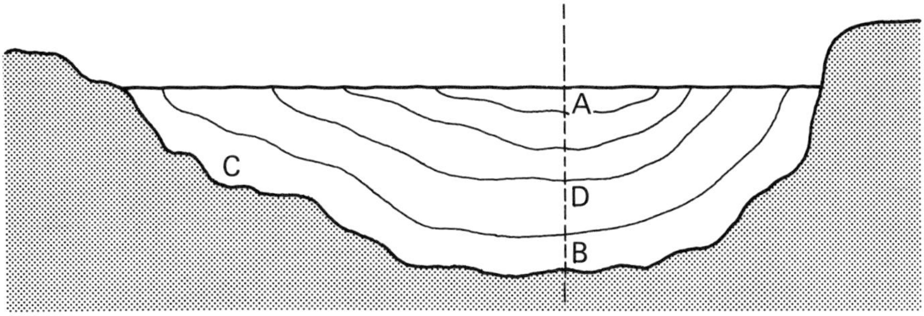

Figure 6–1. Stream Cross Section; Lines of Equal Velocity

(B) Suppose we wanted to estimate the population density of some planktonic organism from the same stream. We could take water samples, concentrate the plankton, and count the number under a microscope. If the stream flow is large and the water is turbulent, it might make little if any difference as to where in the cross-sectional plane we took our samples. For purposes of estimating the population density, this sampling procedure might be very easy. Still, seasonal differences might be profound.

(C) To assay dissolved oxygen in the stream the sampling problem might be as simple as in problem (B), but we would expect that the conditions might vary greatly from one time of day to another. Since the various oxygen sinks are fairly uniform throughout the day in their demand and the input of oxygen by photosynthesis occurs only in daylight, one would expect a fairly obvious diel fluctuation. Here we might choose to sample only at a particular time of day—for example, dawn. Clearly, procedures of sampling will be very important, and in fact one can not exaggerate this aspect of experimental technique; we return to sampling in some detail in a later section.

For the purposes of description, there is a meaningful distinction to be made between continuous and discrete variables. Such measures as counts of objects are obviously discrete, but there is often a point of compromise because it is simply not practical to try to measure a continuous variable continuously. For example, although time is obviously a continuous variable, we rarely measure to closer than the nearest tenth or hundredth of a second. Commonly, we are content to measure to the nearest day or even the nearest year. We know that because of the particulate nature of matter, mass is actually discontinuous, but we tend to think of it as continuous. In a practical sense, then, a variable is regarded as *continuous* if it is measured to decimal places, and as *discrete* if the array of possible values are integers.

For purposes of illustrating several descriptive statistics, we use data from an actual set of observations. Each individual observation, or datum, is the survival time of one organism. The time could be measured in seconds, hours, or years. In this particular case the time happens to be years. There are two groups of organisms, and each group was exposed to a slightly different set of environmental risks. These risks could be exposure to pollutants or any sort of hazard, but in this case the risk was public life. One group consists of the age at death of U. S. presidents (group A); the other consists of age at death of men who served as vice-president but never became president (group B). It should be evident to the reader that the principles are in no sense different from one type of problem than for another.

Group A				*Group B*		
67	54	71	91	80	62	86
90	66	68	79	73	63	76
83	65	58	46	65	68	98
85	77	61	65	51	66	77
73	56	63		68	96	79
81	67	68		70	79	
78	63	58		72	55	
80	71	61		67	66	
68	50	90		54	57	
72	56	63		82	71	

A tabular presentation of the data provides all of the information that we have. With respect to the deceased presidents, it actually constitutes an exhaustive listing of the entire population. These life spans are of course part of some larger universe or population. Perhaps we might consider as a population all the U. S. males who were born from 1732 (the year of George Washington's birth) to 1917 (the year of John F. Kennedy's birth). We might be interested in asking whether or not our two arrays of men—samples of the larger population—showed comparable patterns of longevity. We might wish to compare the two samples and ask whether or not they could reasonably be regarded as coming from the same general population. An important point of rationale comes of such considerations. Clearly, the samples with which we are dealing are not at all similar to the population of all U. S. males. At the time of the revolution, life expectancy of males was approximately thirty-nine years, since many died in infancy or early childhood. The population that might have some relevance here would be all men who survived to such an age that they could have attained presidential office. The youngest ever elected was Teddy Roosevelt at age forty-two. Actually then, we know little about the population from which these data constitute a sample. We can use the sample, however, to try to characterize the entire population. We can note that in group A the minimum longevity was forty-six years, the maximum ninety-one and the range forty-five. We could order, or summarize, the data into frequency distributions:

Group A

40–49	50–59	60–69	70–79	80–89	90–99
1	6	12	8	5	2

Group B

40–49	50–59	60–69	70–79	80–89	90–99
	4	8	8	3	2

Note here that the groupings must be mutually exclusive and all inclusive. No matter what the life span, it will unambiguously fall into one class. This form of presentation is somewhat graphic and gives the reader a feel for the form of the distribution. This can be improved by use of a histogram.

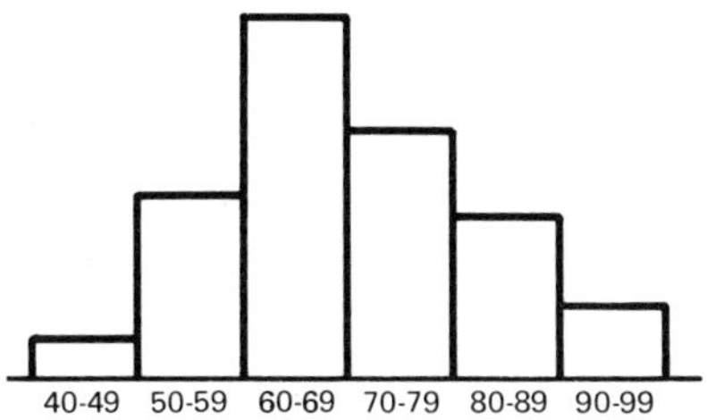

Group A

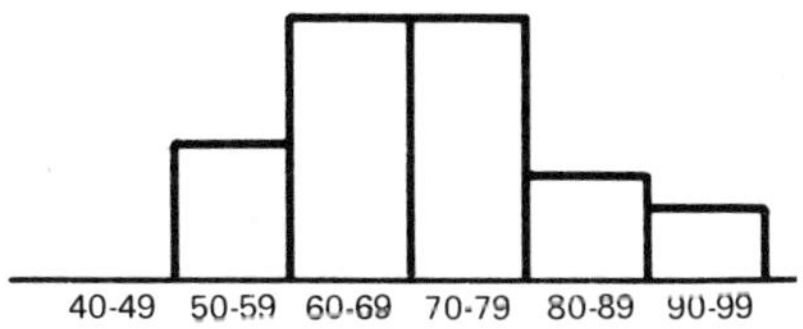

Group B

Clearly the frequency distribution and the histogram do lose some information. The individual data are lost, and the specific individual values are lost. In general, where one has a larger number of data points, the number of classes should be larger—for example, ten to twenty. Here we have so few points that form would tend to be lost if the points were more spread out.

Certain arithmetic deductions from frequency distributions help to describe or characterize a distribution. If we had all the information about a population, we could, of course, give as complete a description as we wished. Because we are usually working from a sample that is only a subset of the totality of possible observations, it is necessary to use some ingenuity to learn as much as we can about the population from which the sample is drawn. The moments of the distribution are most useful in this regard. The most crucial information comes from the first few moments, and here we are primarily concerned with the first two.

The first moment of a distribution locates the "center of gravity" along the real line. Consider a sample of size n taken in some appropriate way from a population. Call the first of these values X_1, the second X_2, the ith X_i, and the nth X_n. We thus have a set of values $X_1, X_2, X_3, \ldots X_i, \ldots X_{n-1}, X_n$. The first moment is

$$m_1 = \frac{X_1 + X_2 \ldots X_n}{n} \tag{6.2}$$

This is the arithmetic mean and is usually written as $\bar{X}$ rather than m_1. It is generally more convenient to use a summation notation

$$\bar{X} = \frac{1}{n} \sum_{i=1}^{n} X_i \tag{6.3}$$

The arithmetic mean is a measure of central tendency—not the only one (as we see later in this section), but one of the most useful.

The second moment corresponds to the moment of inertia about the origin and is defined as

$$m_2 = \frac{1}{n} \sum_{i=1}^{n} X_i^2 \tag{6.4}$$

At this point we encounter two minor problems. First, we are more interested in the moment of inertia about the center of gravity than about the origin. This would be reflective of the spread of the observations about the mean. Second, we are ordinarily dealing with sample data rather than the complete populational information. To cope with the first problem, we shift the distribution center from the mean to the origin by substracting from each observation value (or datum) an amount equal to the sample mean. Thus we would be working with a new set of values, $X_1 - \bar{X}, X_2 - \bar{X}, \ldots X_i - \bar{X}, \ldots X_n - \bar{X}$, and a moment could be obtained:

$$m_2 = \frac{1}{n} \sum_{i=1}^{n} (X_i - \bar{X})^2 \tag{6.5}$$

If this were a population of size n and all possible observational values were included, $\bar{X}$ would be the true population mean (ordinarily written as μ) and this moment would be the populational variance and be written as σ^2. So

$$\sigma^2 = \frac{1}{n} \sum_{i=1}^{n} (X_i - \mu)^2 \tag{6.6}$$

But we are generally not dealing with a whole population, only a sample, and our knowledge is not complete. Thus we do not know the true value of the parameter μ and only have an estimate of it—that is, $\bar{X}$. Because we squared the parenthetic part of the equation, all of the values must of necessity be positive, and this distribution has a peculiar but very useful distribution. Since it is always positive we can err in our estimate only down to the value of 0 on the left, but on the right we can err much more. As a result we tend to underestimate the value of σ^2 more frequently than we overestimate it—but to underestimate it by a relatively smaller amount. If we were to substitute the sample value $\bar{X}$ for μ, we would have a consistent bias in our estimation of σ^2. We can correct for that bias by means of a very simple device and get an unbiased estimate of σ^2—that is, the sample variance s^2. This would be

$$s^2 = \frac{1}{n - 1} \sum_{i=1}^{n} (X_i - \bar{X})^2 \tag{6.7}$$

This is a rather awkward computational form, and it is easier to use the equivalent

$$s^2 = \frac{\sum X_i^2 - \frac{(\Sigma X)^2}{n}}{n - 1} \tag{6.8}$$

Thus the first two moments provide us with unbiased estimators of the location along the real line of the arithmetic mean $\bar{X}$ and the sample variance s^2.

Although each of the higher moments will provide more information about the frequency distribution, we only mention the third and fourth moments and ignore the others. The third moment reflects the degree to which a distribution is skewed or asymmetrical. If there are more observations that fall to the left than to the right of the mean, then the distribution is said to be positively skewed; if there are more to the right of the mean than to the left, then it is said to be negatively skewed.

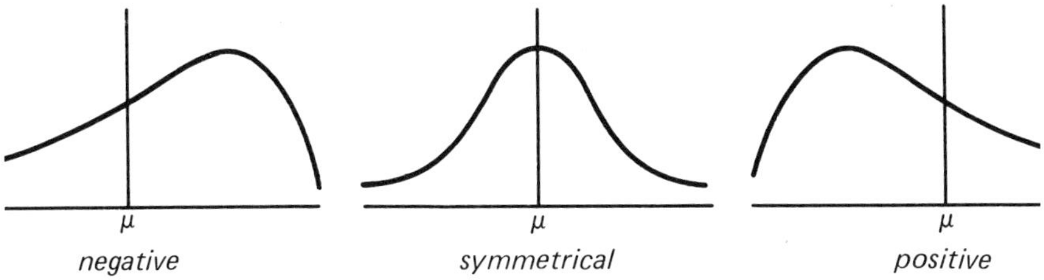

The third moment of a distribution will tend to be near 0 if the distribution is close to symmetrical and will take on the sign of the longer tail of the distribution.

The fourth moment reflects the degree to which the distribution is truncated or chopped off at the tails. This truncation is termed *kurtosis*. One of the theoretical distributions that is very important to us happens to be symmetrical and theoretically tails off to positive and negative infinity. The point of mentioning these two moments—that is, 3 and 4—is that an awareness of asymmetry or kurtosis can alert the user of a potentially serious violation of basic assumptions.

Some other measures of central tendency have special utility. The *median* is that sample value which has half of the observatons larger and half of the observations smaller. In other words, it is the middle value if all observations were ranked by magnitude. In a symmetrical distribution the arithmetic mean and the median will tend to be quite close. Another measure used sometimes with discrete or grouped data is the *mode*. This would be that class value or integer value that is taken the most frequently. The mode is therefore the most popular value. For example, if a population is actually a mixture of two groups, each of the groups may have its own modal value for some measure. As a result, if we were to look at a frequency distribution, we might see these two modes reflected by two peak frequencies. This would not be at all apparent if we only considered the arithmetic mean or the median.

While the variance or its square root, the *standard deviation*, is the usual measure of the spread of observational values, another statistic is occasionally used—that is, the *range*. The range is defined as the spread from the smallest to the largest observation. The only utility of the range is that it reflects the occurrence of "outliers" or extreme values. In some applications it can be useful as a quick and easy way of showing a difference between two samples in the event that the ranges do not overlap.

Characterizing Environmental Data

Our primary concern with empirical environmental data is to put it in a form that can be used in modeling. No model is better than the empirical data that serves as a basis for theoretical projections. The model may be a theoretical construct, but the parametric values should reflect some sort of reality. We need to obtain these estimates in the best possible way and be able to put some sort of qualitative evaluation on them. How good are our estimates? To what degree dare we put trust in them? There are some statistical strategies which permit us to make objective evaluation of the estimates. Here we are very much interested in optimizing our effort in making the best possible parametric estimates for use in the synthesis of theoretical models.

How do we obtain the best possible estimate of a parameter and evaluate the quality of that estimate? The first problem in dealing with empirical information is that of sampling. In general, we seek to make statements about the nature and the parameters of some population on the basis of a sample. A sample consists of some subset of the total universe of possible observations. If the sample is to reflect the true attributes of the population, then it must be representative and sufficient in terms of information to allow us to have faith in our estimate.

The most difficult problem in sampling has to do with the representativeness of the sample. Taking the dissolved oxygen in a river as an example, consider some of the sampling difficulties. We know that from one point of space to another at a single moment of time, there is no reason to hope that all measurements will be alike. Oxygen values at the surface of the river are different from those at the bottom. The aeration in a riffle will result in a difference from that in a pool, and a bed or vegetation in a shallow will result in a difference in oxygen downstream when compared with a deep, slow-moving segment of a stream. Knowing this, what points in space should we choose to take our sample? Were we to choose a point in space should we take samples at various times, and if so, what times? It is clear that at night metabolic oxygen demand will tend to deplete DO, although there will be surface/air interchange and aeration. By day the photosynthetic activity by aquatic vegetation will influence the DO downstream. When should we sample, and where?

This problem stands as one of the most difficult posed for the environmental scientist. The one general solution tends to be to take samples everywhere and all the time—but clearly this is not realistic. A compromise might be to select some points at random, and then at these points to take periodic samples. It is important to realize that this apparently ideal solution would not satisfy those who are concerned about the problem of minimizing the worst conditions, nor would it satisfy those who feel that the most meaningful measure is that which reflects the productivity as measured by the peak diel production.

Both of these points have merit, however, but the problem is one of appreciating what is reflected by a measure obtained in a particular sampling regimen. Persisting in our analysis of a stream, were we to sample repeatedly at a point in space, we might obtain a good reflection of the conditions at this point. Were we to execute similar measurements at another point, we would have a basis for comparison. If we repeated this procedure at enough points, then we might feel that we had done a good job of characterizing the DO conditions in this particular river. We could then be concerned about seasonality and the degree to which we obtained information which would allow us to evaluate the basic mechanisms underlying the overall dynamics of the river.

We do not make any pretense of having "right answers," but what we do have is a strategy that allows us to say a great deal about prevailing conditions in a heterogeneous environment. There are some accidents of statistics that tend to work in our favor. We can use these to help in our task of characterizing natural systems. One example is that sample means, even though they may be means from various and rather peculiar distributions, tend to be normally distributed. The significance of this statement is most profound, and we discuss the significance of the behavior of sample means in considerable detail.

Hypothesis Testing

A formal definition of a statistical hypothesis is any statement about a probability distribution. This is a sound definition, and it actually fits well operationally. For example,

X is a normal random variable

X is a binomial random variable

X is a Poisson random variable

X is a normal random variable with mean $\mu = \mu_0$

X is a Poisson random variable with $\lambda = \lambda_1$

X is a normal random variable with $\mu = \mu_0$ and variance $\sigma = \sigma_0$

Review carefully these statements: The first four are termed *composite hypotheses* because they include many possible hypotheses, each which can completely specify the distribution. The fifth and sixth are simple hypotheses—that is, they not only state the form of the distribution, but they specify the parametric values. The fourth hypothesis is a composite because it specifies mean, but not variance, and the normal distribution depends on both mean and variance. The fifth hypothesis is a simple hypothesis because the Poisson is a single parametric distribution.

The underlying principle of hypothesis testing can be simply stated. We formulate a statement in the form of a statistical hypothesis, and then consider an alternative hypothesis. We choose to reject the hypothesis tested when the alternative is relatively likely. We then design an experiment that will furnish data on which to base a decision. If the outcome is very unlikely under the conditions of the tested hypothesis, it is rejected in favor of the alternative. In a sense we are saying that we disbelieve the hypothesis if an event is seen that is not likely to occur under the assumption of the hypothesis. We think error in posing the hypothesis is more reasonable than is our witnessing a miraculous event. This principle can be shown through a series of illustrations. It is important to note here that in experiments we are able to control conditions to a great extent, but because we are observing a random variable—that is, looking at a sample subject to sampling and observational errors—we might indeed observe a rare event. In such a case it is possible to reject the hypothesis although it is true. Actually, it is also possible to accept the hypothesis when it is actually false, and the alternative is true. The possible outcomes of an experiment are shown here, where the hypothesis (*H*) was made.

When in fact	*Reject H*	*Accept H*
H is true	Error of First Kind	Correct Action
H is false	Correct Action	Error of Second Kind

The following example shows that these two kinds of error might be very different:

A classical case is the production of a drug, which is made by vitiation (semi-killing) of a potentially dangerous, pathogenic organism. The manufacturer tests the drug by injecting samples into a few test animals. If all the test animals are healthy, then the manufacturer concludes that the batch of drug is safe and then bottles it to sell for patient use. If any of the animals get sick, then he concludes that the drug batch is unsafe and then subjects it to further treatment. Consider the possible errors: If the drug is bad, and it is sold and used, then dozens of patients may be infected with the dangerous pathogen. They will get sick, and perhaps die. The victims or their survivors will sue the manufacturer; he may be found negligent and wind up in jail or fall to the violent reprisal of the friends of the victims. If the drug is good, but the manufacturer erroneously thinks it is bad and therefore treats it again, he has wasted time and money. This is a minor problem, since he passes the added costs on to the customer. Obviously these are grossly different consequences of the two sorts of error. In the field of hypothesis testing there is an arbitrary convention—the hypothesis is formulated in such a way that the first kind of error is the worst kind of error. This error of the first kind is always the unjustified rejection of the hypothesis tested. Obviously, we could pose a hypothesis either of two ways. The crucial point is to make this first kind the worst kind of error, and further, to be able to control the probability of making that error. The conventional nomenclature calls for designating the probability of an error of the first kind as (α).

To further specify a set of rules for hypothesis testing, we explore the drug manufacturing process in more detail. Assume that for a given batch

Table 6–1
Probability of Falling in the Critical Region for Various Probabilities of a Drug Dose Causing Disease
(for 1, 3, and 10 test animals)

P	$P(x,1)$	$P(x,3)$	$P(x,10)$
0.0	1.0	1.0000	1.0000
.1	.9	.7290	.3487
.2	.8	.5120	.1074
.3	.7	.3430	.0283
.4	.6	.2160	.0061
.5	.5	.1250	.0010
.6	.4	.0640	.0001
.7	.3	.0270	0.0000
.8	.2	.0080	0.0000
.9	.1	.0010	0.0000
1.0	0.0	0.0000	0.0000

of drug there exists some nonzero probability that a dose-sized sample injected into a test animal or patient could cause disease. If the batch of drug is bad, this probability is relatively high. If a test (experiment) consists of dosing three test animals and examining them for signs of disease, we would conclude that there was no indication that the drug was bad if all three animals appeared healthy. Thus the outcome of the experiment that consisted of zero sick animals would constitute the critical region. Any other observaiton, whether one, two, or three sick animals, would lead us to conclude that the drug was bad and to accept the hypothesis that the batch of drug was bad, because the probability of a dose causing disease is some value greater than 0.

Even if there is appreciable risk, it is possible to miss detecting that the batch is bad. We can indicate the probability of rejecting the hypothesis tested for various levels of alternative by means of the power function. In general, the power function is the probability of falling in a specified critical region as an outcome of a particular test experiment for all of the possible simple hypotheses, including the hypothesis tested and all of the alternatives. Here we calculate the probability of falling in the critical region (observing no sick test animals) for various possible probabilities of a drug dose causing disease. We repeat this for different experiments consisting of one, three, and ten test animals.

In table 6–1 the column headed p is the probability that any given dose will cause disease in any given test animal. The column headed $P(x,1)$ gives the corresponding probability of falling in the critical region that consists of "no sick animals" and so rejecting the hypothesis that the drug is bad if only one test animal is used. The column headed $P(x,3)$ gives the corresponding probability of falling in the critical region and so rejecting the hypothesis when three test animals are used. The column headed $P(x,10)$ gives the corresponding probability when ten test animals are used instead of three.

The power function tells a great deal about the possible performance of a test. Note here that the event of falling in the critical region—that is, rejecting the hypothesis tested when the drug is actually bad—bears the likelihood of grave consequences. If the drug batch is good one is certain of taking the correct actions; rejecting the tested hypothesis simply because a good batch cannot cause disease in test animals or in patients. The problem is to not fall in the critical region when the drug can cause disease, even if the probability is quite small. A graphic representation of table 6–1 illustrates how quickly the probability of an error of the first type drops off with increasing sample size (but the probability of such an error does persist when the probability of a dose causing disease is quite close to 0).

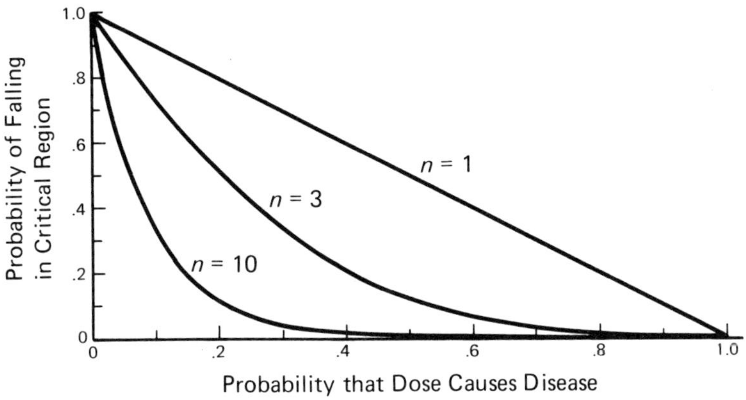

It is not possible to eliminate this possibility, but the probability can be minimized by maximizing the sample size.

Suppose that we have theoretical grounds for predicting that some environmental parameter will have a true value of μ. This could be the result of a purely abstract model, or one for which we have sound empirical information concerning underlying parameters of the model. Further, we assume that single observations of parametric values are distributed as normal with some variance σ^2. Under these conditions we could take a sample of n observations and obtain a sample mean $\bar{X}$ that would be an unbiased estimate of μ and would have a variance $\sigma_{\bar{x}}^2$ which would be an unbiased estimator of σ^2/n. It would be convenient to make the transformation

$$Z = \frac{\bar{X} - \mu_0}{\sigma/\sqrt{n}}$$

This has a normal distribution with mean 0 and variance 1. Table A–1 in the appendix can be used to determine the probability of observing values of Z in various segments of the real line. Because μ_0 is our hypothetical value for the mean μ, we would expect that if we are correct in assuming $\mu_0 = \mu$, then the values of Z which we obtain from observations will tend to be close to 0.

Should we observe a value of Z much different from 0—either a large positive or a large negative value—we would either be witnessing a very likely event or else we would have reason to doubt the truth of our hypothesis. If the value of Z is in an extremely unlikely range, we would feel that our hypothesis has been essentially disproved and want to reject it. On the other hand, if Z is close to 0, then we would feel that the sample value was consistent with the hypothesized mean, and we would choose to accept our tested hypothesis. For practical purposes it is convenient to be rather formal in setting up and testing hypotheses. Follow through these steps:

Express the statistical hypothesis to be tested and specify the alternative—for example,

$$H: \quad \mu = \mu_0$$

$$\bar{H}: \quad \mu \neq \mu_0$$

Specify the test statistic. Here we have chosen

$$Z = \frac{\bar{X} - \mu_0}{\sigma/\sqrt{n}}$$

Determine the distribution of the test statistic under the condition that the hypothesis is true. Here Z is normal with mean 0 and variance 1.

Specify the level of significance α—that is, the probability of unjustifiably rejecting the hypothesis tested.

Determine the array of values of the test statistic for which you would want to reject the hypothesis tested. Here we want to reject the hypothesis if $|Z|$ is large. Such a critical region would be one consisting of two zones such that the probability of falling into either of them is $\alpha/2$ when H is true.

Compute the value of Z based on our known value of σ^2, the sample size n, the hypothesized mean μ_0, and the observed value of $\bar{X}$. Compare this value with the critical region.

Reject H if Z falls in the critical region. Accept H if Z does not fall in the critical region.

Following this procedure routinely, one will consistently make optimal use of data. One point of strategy deserves special mention—the choice of a critical region. The size of a critical region—that is, the probability of unjustifiably rejecting the hypothesis tested—can usually be fixed at or below the arbitrary level α. But some critical regions of size α are better than others in the sense of the resulting probability of detecting the falsity of an hypothesis.

Example 6.1. If we were testing whether or not an aeration device improved oxygen conditions in a lake, we might only be interested in detecting improvement. We would test a null hypothesis that states that the machine makes no improvement. This could be stated as mean dissolved oxygen with treatment (μ_1) equals mean dissolved oxygen without treatment (μ_2). Assume we have sample means and variances. If the sample sizes are the same ($n_1 = n_2$) and the variances are unknown but equal ($\sigma_1^2 = \sigma_2^2$), then the appropriate test statistic would be

$$t = \frac{(\bar{X}_1 - \mu_1) - (\bar{X}_2 - \mu_2)}{S_{\bar{X}_1 - \bar{X}_2}}$$

Since we are hypothesizing equality of the population means,

$$t = \frac{(\bar{X}_1 - \bar{X}_2) - (\mu_1 - \mu_2)}{S_{\bar{X}_1 - \bar{X}_2}} = \frac{\bar{X}_1 - \bar{X}_2}{S_{\bar{X}_1 - \bar{X}_2}}$$

The estimate of the standard deviation of the differencc between the means in general can be obtained from

$$S\sqrt{\left(\frac{1}{n_1} + \frac{1}{n_2}\right)}$$

where

$$S^2 = \frac{\sum X_{1i}^2 - \dfrac{(\Sigma X_{1i})^2}{n_1} + \sum X_{2i}^2 - \dfrac{(\Sigma X_{2i})^2}{n_2}}{n_1 + n_2 - 2}$$

$$= \frac{(n_1 - 1)S_1^2 + (N_2 - 1)S_2^2}{n_1 + n_2 - 2}$$

Where the numbers are the same as stated earlier, this reduces to

$$S^2 = \frac{S_1^2 + S_2^2}{n}$$

Now we obtain the estimate of the variance of the differences between the means

$$S^2_{\bar{X}_1 - \bar{X}_2} = \frac{S_1^2 + S_2^2}{n}$$

So

$$S_{\bar{X}_1 - \bar{X}_2} = \sqrt{\frac{(2S^2)}{n}} = \sqrt{\frac{(S_1^2 + S_2^2)}{n}}$$

The test statistic t is distributed as a t with $2n - 2$ degrees of freedom. To develop a rationale of selecting critical regions, we assume that we have 9 degrees of freedom and $\alpha = .05$.

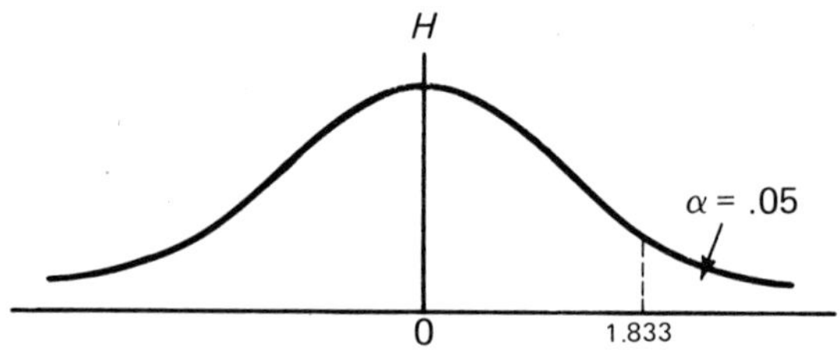

Here we would reject H if we observe a value of t greater than 2.262. Since we would like to detect a good aerator, this is quite appropriate.

We could, however, have a very different critical region with the same α—for example,

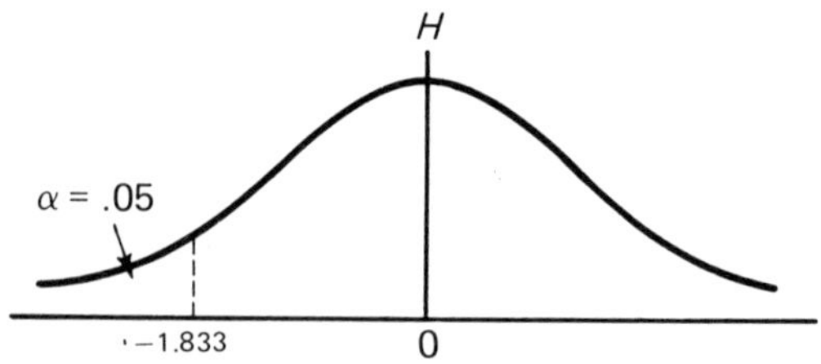

Here we would be very unlikely to reject H if the aerator was actually good, and in fact only be successful in detecting those which worsened the conditions. This is an obviously inappropriate critical region:

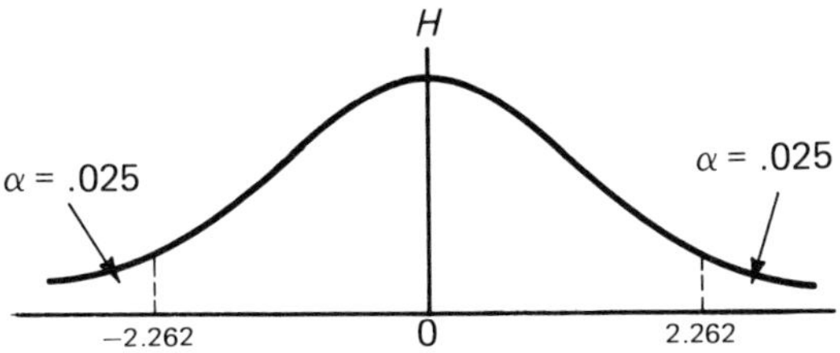

Here, too, we meet the $\alpha = .05$ requirement. The critical region has two parts. The area to the right would detect good aerators, but the area to the left is wasted.

In general, then, the choice of a critical region depends on the alternative hypothesis. We can always improve the probability of detecting the falsity of the hypothesis tested by taking a larger critical region either by increasing α or alternatively by taking larger samples.

The Normal Distribution

One of the fortunate events that makes life easier for those who would apply statistical techniques to sample information is the fact that sample means tend to have a very nice distribution. Keep in mind that sample statistics such as the mean ($\bar{X}$) and variance S^2 are themselves random variables. The sample mean is an unbiased estimator of the true population mean μ and tends to have what is termed a *normal* distribution.

The normal distribution is a distribution that is symmetrical, centered at some parametric value of the mean (μ), and is completely determined by the two parameters μ and σ^2. It is a continuous distribution and can be approached by some very important and useful discontinuous distributions that have theoretical bases which are easy to accept. A derivation of the normal distribution is beyond the scope of this work, and we refer the reader to the many fine works on theoretical statistics. The normal distribution is represented by the familiar bell-shaped curve:

$$f(x) = \frac{1}{\sigma\sqrt{(2\pi)}} \exp\left[-\frac{1}{2}\left(\frac{X - \mu}{\sigma}\right)^2\right] \qquad (6.9)$$

where σ is the parametric value of the population standard deviation
μ is the parametric value of the population mean

Figure 6–2 illustrates a set of possible curves of $f(x)$, all for the same value of μ, but with different values of σ. Here each of the curves has the same value of the mean or first moment. The values of σ reflect the sec-

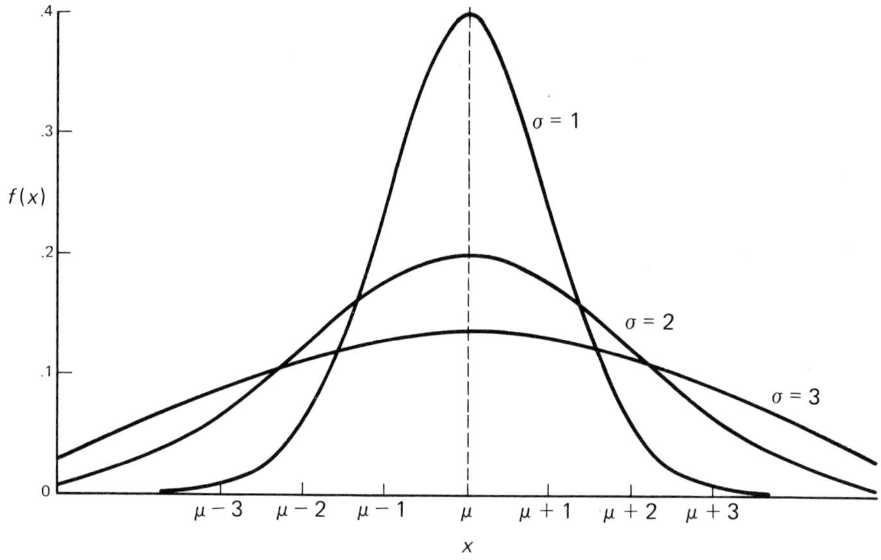

Figure 6–2. Normal Distribution; Constant μ, Variable σ

ond moment of the distributions, and in each case the area under the curve is the same—unit area, for example. The mean could—at least theoretically—take on any value along the real line, and it is possible for σ to take on any positive value along the real line also. Thus there are many possible locations and shapes of the normal distribution. Fortunately, it is so easy to make such necessary transformations that a single graphic or tabular representation of $f(x)$ permits us to determine areas under the normal curve for any values of μ and σ. As an example we can define the standard normal Z as

$$Z = \frac{x - \mu}{\sigma} \tag{6.10}$$

where μ is the mean and σ is the standard deviation of the normal distribution of interest. It can be seen that by subtracting μ from all of the values of x, the new distribution is centered at 0. By dividing all values by σ, the new variance—and hence the new standard deviation—is unity. Thus Z is distributed with mean 0 and standard deviation 1 and is usually written as $N(0,1)$.

Areas under the standard normal curve $P(z)$ are presented in table A–1 in the appendix, and the use of such a table can be illustrated by the following example:

Example 6.2. Suppose x is a normal random variable, with $\mu = 150$ and $\sigma = 10$. We are interested in determining the probability of observing a value of X between 120 and 140. Figure 6–3 illustrates the relationship between the distributions of X and Z.

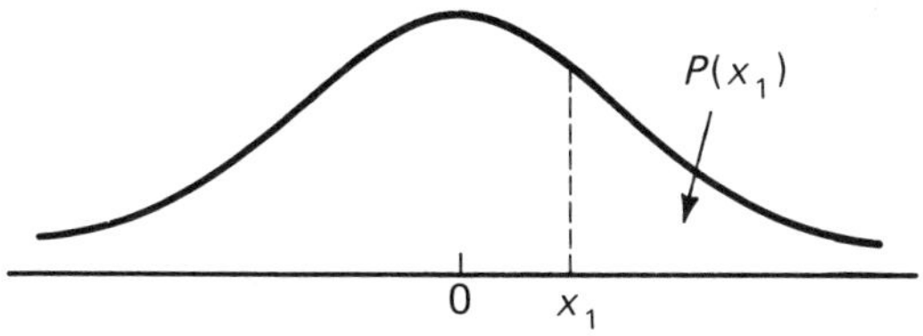

The area in which we are interested is

$$\int_{-3}^{-.6} Z\,dZ$$

Table A–1 in the appendix is set up to read

$$\int_{z_1}^{\infty} Z\,dZ$$

Because the normal distribution is symmetrical, we can write

$$\int_{-3}^{-.6} Z\,dZ = \int_{.6}^{3} Z\,dZ = \int_{.6}^{\infty} Z\,dZ - \int_{3}^{\infty} Z\,dZ$$

From the table

$$\int_{-3}^{-.6} Z\,dZ = \int_{.6}^{3} Z\,dZ = .2743 - .0013 = .2730$$

The probability of observing a value of X between 120 and 144 when X is $N(150,10)$ therefore is .2730.

Texts on statistics typically put much stress on the theoretical aspects of the normal distribution. Part of this is because of the fact that we see distributions approaching the normal in many sorts of natural and experimental situations. Another reason is that sample means tend to approach normality. Sample means tend to be distributed as normal with mean equal to μ and variance σ^2/n where n is the sample size.

By way of illustration we use a series of means, in each case totaling 100, but obtained from samples of various sizes. The population being sampled is the set of random digits, 0 to 9, generated by a computer program. Clearly the real distribution of random digits should be far from normal. Since each digit supposedly has the same probability of appearing as every other digit, the distribution should be uniform, or flat, and range

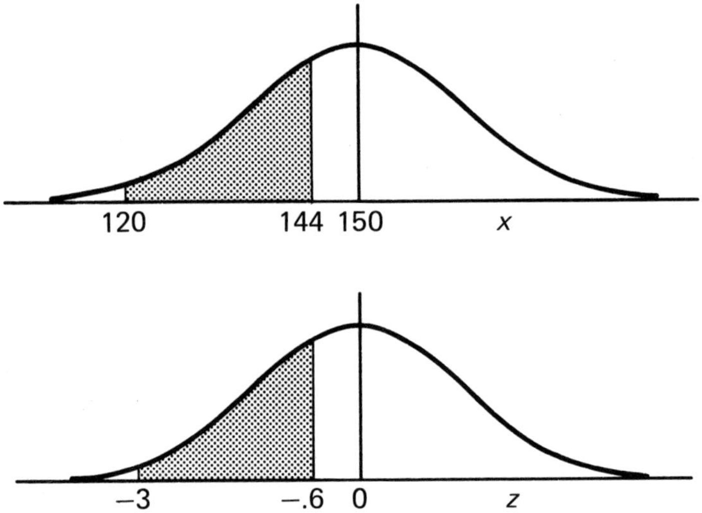

Figure 6–3. Normal between Two Distributions

from 0 to 9. Figure 6–4 shows the result of a little experiment in sampling, and for each set of 100 means a mean of sample means $\bar{\bar{X}}$ is presented along with the standard deviation of the sample means. On theoretical grounds the mean μ should be equal to 4.5 and the population variance $\sigma^2 = 9.1677$. For each of six computer runs we present the actual frequency distribution of the 100 sample means, the sample size, and the mean of sample means $\bar{\bar{X}}$. Also, the actual standard deviation of the sample means $S_{\bar{x}}$ is presented with the theoretical standard deviation of the sample means $\sigma_{\bar{x}}$. Here the theoretical value of the standard deviation of the sample means is $\sigma/\sqrt{n}$. The figure shows clearly that there is a trend toward a tightly distributed normal distribution of sample means even though the parent distribution is far from normal.

The fact that means tend to normality is important in more than a theoretical sense, because it is this tendency that makes it possible to manage our observational errors and to assess their importance.

Binomial Distribution

The binomial distribution is based on a series of n trials, each of which has only two possible outcomes—success or failure. For example, assume the probability of success on each trial is p, and hence the probability of failure is q where $q = 1 - p$. We define the binomial distribution in terms of the probability law for the number of successes X in a series of n trials.

We ask the probability that X will take on the value k where k is any integer from 0 to n. In a series of n trials we can have k successes occur in C_n^k ways. Here

$$C_n^k = \frac{n!}{k!(n - k)!}$$

It is necessary to note here that $0! = 1$.

The probability of any particular array of k successes in n trials is the same for all of those enumerated by C_n^k. For example, if n were 5, and k were 2, then each of the following outcomes could be expected to occur with equal frequency—for example,

SSFFF	SFSFF	SFFSF	SFFFS	FSSFF
FSFSF	FSFFS	FFSSF	FFSFS	FFFSS

where S = success

F = failure

Since the probability of success is the same on every trial, we can specify the probability of any one series—for example, the first could be $p \times p \times q \times q \times q$. Because the various series of outcomes are mutually exclusive, we need only count the number of series that have the requisite number of successes and multiply the count by the probability of that number of successes. The probability that the number of successes X is k in a series of n independent trials, where the probability of success on each trial is p, can be written as

$$P(x{=}k \quad n{,}p) = C_n^k p^k q^{n-k}$$

For our example,

$$P(x{=}2 \quad 5{,}p) = C_5^2 p^2 q^{5-2}$$

To continue the example, suppose that the probability of success p were equal to .20, and that we are seeking the probability that there are two successes in a sequence of five trials:

$$\begin{aligned} P(x{=}2 \quad 5,.2) &= C_5^2(.2^2)(.8^3) \\ &= \frac{5!}{2!3!}(.2^2)(.8^3) \\ &= \frac{5 \times 4 \times 3 \times 2 \times 1}{2 \times 1 \times 3 \times 2 \times 1}(.04)(.512) \\ &= 10(.02048) \\ &= .2048 \end{aligned}$$

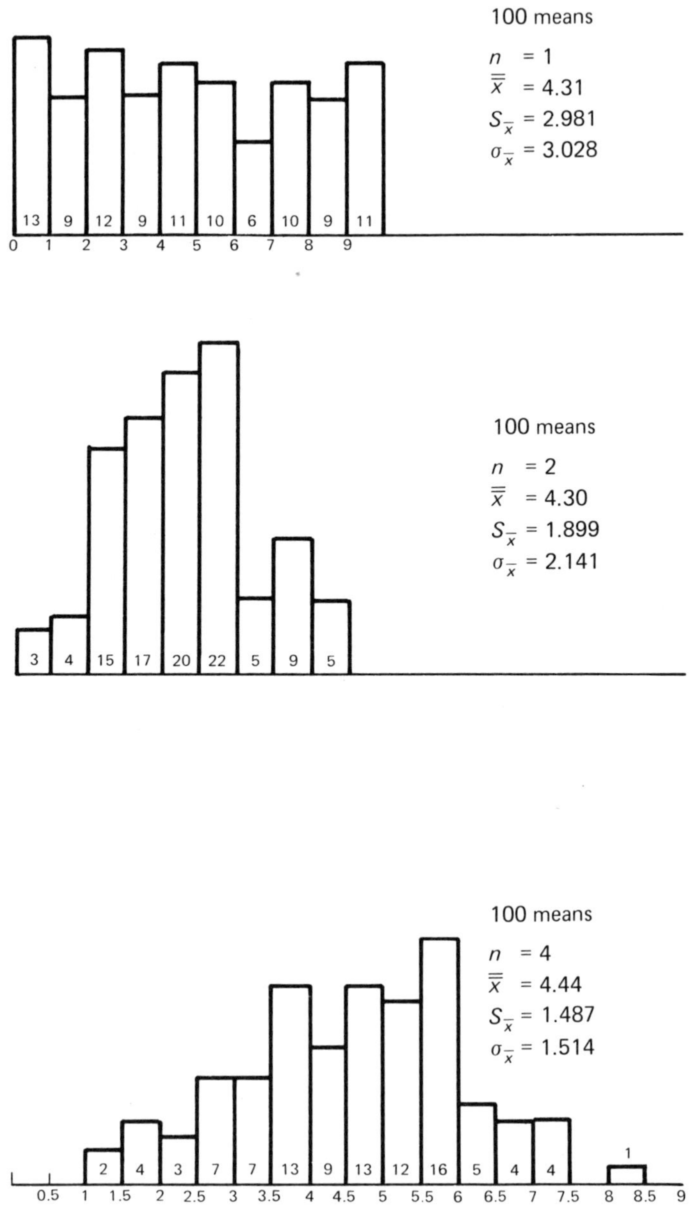

Figure 6–4. Distribution of Means

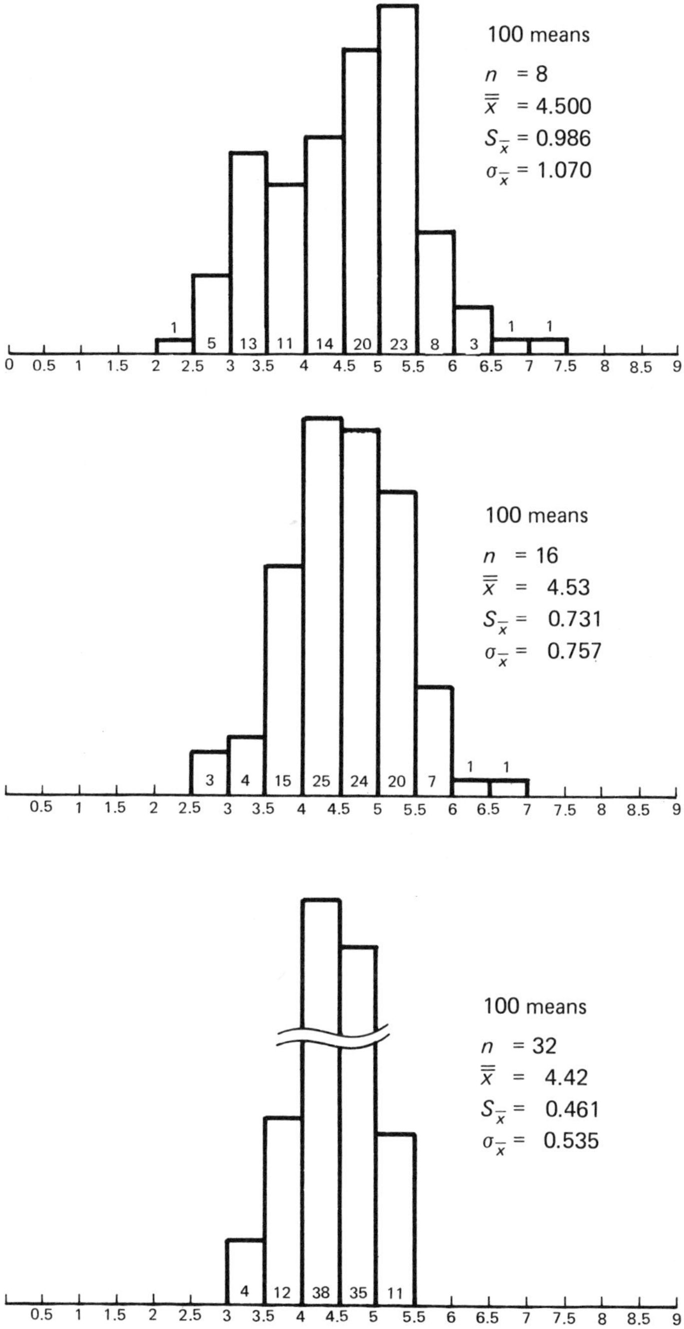

Figure 6–4 (continued)

Similarly, one could compute the probability that x would take on each of the possible values from 0 to n. This would yield a theoretical frequency distribution. This frequency distribution, in general, has a mean equal to np and variance to npq. The binomial frequency distribution, then, is completely determined by the parameters n and p, since $q = 1 - p$. The shape of the binomial distribution will be profoundly influenced by n and p. Figure 6–5 illustrates three quite different shapes brought about by varying p. In each case the number of trials is the same—that is, twenty. For one distribution $p = .01$, for another $p = .1$, and for the third $p = .4$.

Recall that for the normal distribution the form was always the same—that is, symmetrical, defined from minus infinity to plus infinity, and centered about the mean μ. The binomial distribution, however, is discontinuous, bounded by 0 and n, and may be very asymmetrical. It is interesting and important that if n is large, and if p is close to neither 0 nor 1, the binomial tends toward the normal as n approaches infinity. It is often practical to use the normal distribution to estimate the probability of X falling between specified values by using the normal approximation to the binomial. The reader will note that n need not become very large before the computation of exact values of the probability that $x = k$ becomes quite tedious.

The Student's *t* Distribution

The standard normal distribution

$$Z = \frac{\bar{X} - \mu}{\sigma/\sqrt{n}} \tag{6.11}$$

is a random variable because of the term $\bar{X}$. Both μ and σ are parametric values of some normal distribution, and n is the size of the sample from which $\bar{X}$ was calculated. The distribution has mean 0 and standard deviation 1. If we did not know the true parametric value of σ and instead had to estimate it by S, then

$$t = \frac{\bar{X} - \mu}{S/\sqrt{n}} \tag{6.12}$$

should have a more or less similar distribution. Here we have two random variables—that is, $\bar{X}$ and S. The larger the sample size the better we would expect our estimate of S to be, but if n were small we might expect to encounter appreciable error in S, the estimate of σ.

The distribution of t was first worked out by Gossett, who published

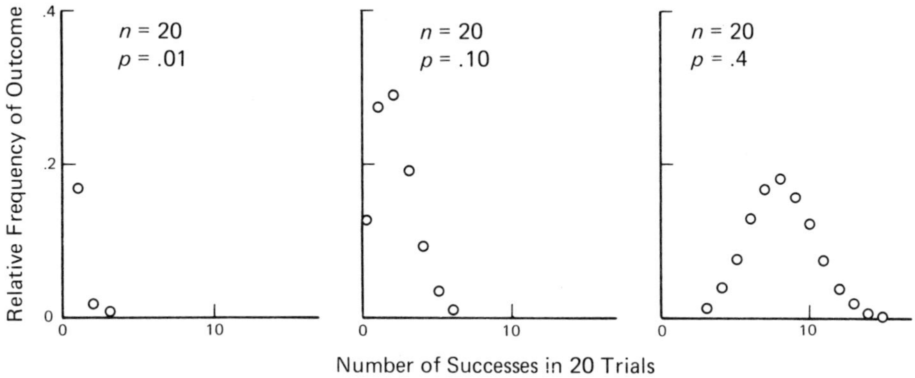

Figure 6–5. Binomial Distributions as Influenced by p

under the pseudonym Student. Table A–2 in the appendix presents some selected values of the Student's t variable. The first column is the number of degrees of freedom (ν), which is generally the sample size minus 1. There are some important exceptions to this generalization, and these are brought to the reader's attention as we encounter them. The various columns are headed with values representing the probabilities numerically exceeding the entry value. The probabilities thus are the sums of the probabilities in the two tails of the distribution.

To illustrate the use of the Student's t distribution, we consider a random variable X that is normally distributed with mean μ and variance σ^2. We have taken a sample of size n and obtained as estimate of σ—that is, S. If $n = 7$ and $S = 4.26$, and assuming that μ is really 36, find values of $\bar{X}$ such that more disparate mean values would be found only 10% of the time:

$$t = \frac{\bar{X} - \mu}{S/-\sqrt{n}} = \frac{\bar{X} - 36}{4.26/-\sqrt{7}}$$

For 6 degrees of freedom, $t = 1.9432$. So

$$1.9432 = \frac{\bar{X} - 36}{4.26/-\sqrt{7}}$$

Solving for $\bar{X}$,

$$\bar{X} = 36 \pm 1.9432(4.26/\sqrt{7})$$

$$\bar{X} = 36 \pm 3.12$$

χ^2 Distribution

The chi-square (χ^2) distribution can be most easily introduced by means of an example. Going back to the information used to illustrate the tendency of sample means to be normally distributed, we list here computer generated pseudo-random digits. The first set of 100 random digits occurred as follows:

Cells	*0*	*1*	*2*	*3*	*4*	*5*	*6*	*7*	*8*	*9*
observed	13	9	12	9	11	10	6	10	9	11
expected	10	10	10	10	10	10	10	10	10	10

We expect each digit to occur with the same frequency if our random number generating program is working as planned so that we do have theoretical or expected frequencies. A way to test whether or not this expectation is met is to examine the magnitude of this measure:

$$\chi^2 = \sum_{i=1}^{h} \frac{(O_i - E_i)^2}{E_i} \tag{6.13}$$

Here O_i represents the observed frequency of occurrence in the ith digit, while E_i is the expected frequency of the ith digit. For this case,

$$\chi^2 = \frac{(13 - 10)^2}{10} + \frac{(9 - 10)^2}{10} + \frac{(12 - 10)^2}{10} + \frac{(0 - 10)^2}{10} + \frac{(10 - 10)^2}{10}$$

$$+ \frac{(10 - 10)^2}{10} + \frac{(6 - 10)^2}{10} + \frac{(10 - 10)^2}{10} + \frac{(9 - 10)^2}{10} + \frac{11 - 10)^2}{10}$$

$$= \frac{34}{10} = 3.4$$

Clearly, the less the expected and observed values conform, the larger the measure will be. Although for 100 occurrences partitioned into ten cells there are only a discrete number of possible outcomes—and hence, χ^2 must theoretically then have a discrete distribution—it is more convenient to use a continuous approximation. This distribution is presented in table A–3 in the appendix. It depends only on the number of degrees of freedom, a point which requires some explanation.

In our illustration there were 100 events to be partitioned into ten cells. Once nine of the cells contain their allocated number of events, the tenth is determined. Nine are free to vary, but the last is determined because the total number of events is set as is the number of cells; therefore there are 9 degrees of freedom. In general, the number of degrees of freedom is the number of cells free to vary.

Note that this use of the χ^2 statistic is limited to enumeration or counting statistics. Further, the approximation is only good if the expected number of occurrences in all cells is reasonably large. This is because the expected number is a denominator, and a small deviation from expectation can be grossly inflated. Generally, one seeks to avoid cells with expectations less than 5. It is quite appropriate to combine cells to avoid this difficulty.

Example 6.3. Using the standard procedure of testing hypotheses with the preceding illustration,

$$H: \quad P_0 = P_1 = P_9$$

$$\bar{H}: \quad P_i \neq k \text{ for all } i$$

(Probability of each digit is the same; $P_i = k$ for all i).

Specify test statistic:

$$\chi^2 = \sum \frac{(O_i - E_i)^2}{E_i}$$

is distributed as χ^2 with 9 degrees of freedom.

Set $\alpha = .05$ (arbitrary).

Specify critical region; if χ^2 exceeds 16.919, then reject H.

Calculate

$$\sum \frac{(O_i - E_i)^2}{E_i} = 3.4$$

Compare with critical region.

Since 3.4 is less than 16.919, we have no basis for rejecting H. Therefore, we accept the hypothesis of uniform probabilities for the occurrence of the various digits.

The χ^2 distribution is extremely useful in a wide variety of situations. Some of these are limited only by the ingenuity of the experimenter. This device permits one to design experiments in such a way as to avoid troublesome frequency distributions. The χ^2 tests may not have as much power as several others, but they are relatively independent of the nature of the underlying distributions. Very important in this regard is that we can do experiments where the outcome is qualitative or subjective.

As an hypothetical example suppose we have four alternative designs for a development, and we ask two different groups of people which they prefer. One group consists of builders, while the other group consists of potential purchasers:

Designs	*A*	*B*	*C*	*D*	*Totals*
buyers	34	18	56	40	148
builders	17	6	19	22	64
Totals	51	24	75	62	212

Here let us ask whether buyers and builders are alike in their preferences:

$$H\colon\ P_{A1} = P_{A2},\ P_{B1} = P_{B2},\ P_{C1} = P_{C2},\ P_{D1} = P_{D2}$$

$$\bar{H}\colon\ \text{not so}$$

Specify test statistic:

$$\chi^2 = \frac{(O_{ij} - E_{ij})^2}{E_{ij}}$$

Under H, this is distribution of χ^2 with 6 degrees of freedom. The reason there are 6 degrees of freedom is that the number of buyers is fixed as is the number of builders. In each case there are three designs which can vary, but then the number for the fourth design is fixed by the total number.

Set $\alpha = .05$ (arbitrary).

Specify critical region; if χ^2 exceeds 12.592, reject H.

Calculate

		A	B	C	D	
Buyers	observed	34	18	56	40	148
	expected	35.60	16.75	52.36	43.28	
Builders	observed	17	6	19	22	64
	expected	15.40	7.25	22.64	18.72	
		51	24	75	62	212

$$\sum \frac{(34 - 35.60)^2}{35.60} = 2.2085$$

Compare with critical region: 2.2 is less than 12.6, so accept H.

Here, buyers and builders agree in their judgments. Clearly the information underlying preferences might be radically different, but the final outcome shows similarity of preference.

Another use of χ^2 is to test for goodness of fit—that is, in general, to determine how well some observational data conform to a theoretical distribution. As an example, here we introduce another interesting and very useful distribution—the Poisson.

A very general situation in environmental processes is that in which there is some unknown small probability of an event occurring in a short interval of time. For example, we might have a monitoring device (sampling from a body of water) conducting a continuous but very small volume of water through a counter which indicates the number of particles larger than some arbitrary size passing in an interval of time. The observer can set the particle size and can also set the duration of the time interval. If the interval is very short, then most of the intervals might have no particles pass, some would have one, few have two, and very few have three or more. This type of process is similar to that of the occurrence of ionizing radiation in a small defined space or the decay process of a radioactive substance. Even the occurrence of individuals in one-, two-, or three-dimensional space can be seen to fit this general scheme.

This general phenomenon is termed a *Poisson process* and can be simply defined. If the probability of an event in unit time is p and the length of a time interval is t units, then there is a rate of occurrence of events $p \times t$; this is constant, and is called λ. Now if we were interested in the frequency with which we observed zero events in an observation interval, then we could express the relative frequency of intervals with no events as

$$\frac{\lambda^0 e^{-\lambda}}{0!} = e^{-\lambda} \tag{6.14}$$

For 1, 2, 3, . . . n event, we would have

$$\frac{\lambda^1 e^{-\lambda}}{1!}, \quad \frac{\lambda^2 e^{-\lambda}}{2!}, \quad \frac{\lambda^3 e^{-\lambda}}{3!} \ldots \frac{\lambda^n e^{-\lambda}}{n!} \tag{6.15}$$

or, in general,

$$P\{k \text{ events in an interval}\} = \frac{\lambda^k e^{-\lambda}}{k!} \tag{6.16}$$

It is both important and convenient that for the Poisson distribution, the mean is λ and the variance is also λ. Thus this distribution is entirely determined by a single parameter; this parameter reflects the rate of occurrence or the density of points in space.

By way of illustration consider an example in which we might on a priori grounds suspect that conditions are such that Poisson conditions would prevail. We have a set of sixty-two bottom samples taken with a dredge and ask the frequency of occurrence of zero, one, two, etc. snails in the samples. If the snails are randomly distributed so that the probability that any given snail will fall into any specified sample, then the density of snails alone should determine a Poisson distribution. From the observed mean density we can then determine a hypothetical distribution.

	Number of Snails in Sample							
	0	1	2	3	4	5	6	7
observed	27	14	7	8	0	2	3	1
expected	15.2	21.4	15.0	7.0	2.5	.7	.2	0

The observed "J"-shaped distribution might well lead us to feel that this distribution is Poisson. Nevertheless, we go through our standard testing procedure:

H: X is a Poisson random variable with λ the mean frequency observed.

$\bar{H}$: X is not a Poisson random variable.

As a test statistic, we use

$$\chi^2 = \frac{(\text{observed} - \text{expected})^2}{\text{expected}}$$

Since we can determine $\bar{X}$ from the observations, we can calculate the expected frequency of occurrence in each cell. Pooling the values in the tail of the distribution to avoid cells with expectation of less than 5, we obtain

	Number of Snails in Sample			
	0	1	2	3 or more
observed	27	14	7	14
expected	15.2	21.4	15	10.4

Arbitrarily make $\alpha = .05$.

Referring to table A–3 in the appendix, we find that under our hypothesis we would expect to observe a value of χ^2 greater than 7.81, 5% of the time. This corresponds to a critical region of size $\alpha = .05$ with 3 degrees of freedom because we have four cells (but when the contents of three of them are determined then the fourth is determined).

To compute the numerical value of the χ^2 statistic,

$$\chi^2 = \frac{(27 - 15.2)^2}{15.2} + \frac{(14 - 21.4)^2}{21.4} + \frac{(7 - 15)^2}{15} + \frac{(14 - 10.4)^2}{10.4}$$

$$= 9.16 + 2.56 + 4.27 + 1.25$$

$$= 17.24$$

Since 17.24 exceeds 7.81, we reject the hypothesis tested. In this case snails only live where there is a certain type of bottom vegetation and are absent where the bottom is barren. As a result their density is fairly high in vegetated areas—considerably higher than the average density. Clearly the assumption of random distribution is violated here, and we properly are led to reject H.

Confidence Interval Estimation

Because the normal distribution is determined by μ and σ, the estimation needed to characterize such a distribution is simple. We encountered the unbiased estimators for μ and σ earlier that is, $\bar{X}$ and S. These estimates give single values which, as sampling experiments increase, will tend to give increasingly better estimates of the unknown true values. The more representative and the larger the sample size, the better the estimate will tend to be. In addition to such point estimates, it is possible to obtain interval estimates that can be made in such a manner as to reflect the quality of the estimate and hence permit the ascription of an arbitrary degree of confidence in the estimate. Such an estimate is termed a *confidence interval* estimate. The definition of a confidence interval is very important for an accurate interpretation: A confidence interval is an interval so constructed that it is one of a population of such intervals which would with some specified probability include the unknown population parametric value.

This does not mean that any particular interval will include the true value. If infinitely many intervals were chosen in the specified manner, some proportion of those intervals would include the true value. We illustrate this point with a synthetic but realistic example.

Example 6.4. In a marine bay a series of salinity observations are made. The common measure of salinity groups all of the salts and is expressed as grams of salts per kilogram of sea water. Salinity is usually expressed as parts per thousand and written with the symbol $^{o}/_{oo}$.

Assume that we know the standard deviation of salinity values in other coastal bays to be $\sigma = 1.40$. On the basis of twenty salinity observations, we estimate the mean as $\bar{X} = 36.000$. The sample mean should be distributed as a normal, with mean μ and a standard deviation:

$$\sigma = \frac{\sigma}{\sqrt{n}} = \frac{1.40}{\sqrt{20}} = .313$$

and

$$Z = \frac{\bar{X} - \mu_0}{\sigma_{\bar{X}}}$$

which is distributed as a normal random variable, with mean 0 and standard deviation 1.

Suppose we wanted to determine a 95% confidence interval for μ. Find an interval that is one of a population of intervals, 95% of which will include the unknown true parametric value μ. Referring to table A–1 in the appendix for the distribution of Z it can be seen that we have 2.5% of the area below -1.96 and 2.5% of the area above $+1.96$. Since the quantity

$$\frac{X - \mu}{\sigma_{\bar{X}}}$$

is distributed as Z, it follows that if

$$P\{-1.96 < Z < 1.96\} = .95$$

then

$$P\{-1.96 < \frac{\bar{X} - \mu}{\sigma_{\bar{N}}} < 1.96\} = .95$$

Solving the inequalities for μ,

$$\bar{X} - 1.96\sigma_{\bar{X}} < \bar{X} + 1.96\sigma_{\bar{X}}$$

For our example on salinity estimation,

$$36.000 - 1.960(.313) < \mu < 36.000 + 1.960(.313)$$

$$35.387 < \mu < 36.614$$

The 95% confidence interval (95% CI) is written as

$$35.387 \text{ —— } 36.614$$

In general we do not know the true parametric values. For example, suppose that instead of 1.4 being the parametric value of σ, it was a sample standard deviation S that was obtained from the same sample of size 20 as the mean. Therefore,

$$\frac{\bar{X} - \mu}{S_{\bar{X}}}$$

would not be distributed as Z normal (0,1), but rather would be distributed as t (19 degrees of freedom).

Referring to table A–4 in the appendix for 19 degrees of freedom or

$$P\{-2.093 < t(19 \text{ degrees of freedom}) < 2.093\} = .95$$

then

$$P\{-2.093 < \frac{\bar{X} - \mu}{S_{\bar{X}}} < 2.093\} = .95$$

Again solving the inequalities,

$$\bar{X} - 2.093(S_{\bar{X}}) < \mu < \bar{X} + 2.093(S_{\bar{X}})$$

For the problem on salinity estimates,

$$36.000 - 2.093(.313) < \mu < 36.000 + 2.093(.313)$$

$$35.345 < \mu < 36.655$$

The 95% CI for μ is

$$35.345 \text{ —— } 36.655$$

Comparing the two confidence intervals that we have obtained, note that the first, where we assumed that we knew the true parametric value of σ, is a little bit shorter. This is a reflection of the difference between the Z distribution and the Student's t distribution. The t distribution has the added variability due to the fact that not only is the sampling variability of estimating the mean present, but also the sampling variability due to estimating the variance. Look back to table A–4 in the appendix and note the importance of sample size. It enters the denominator of the standard error calculation, and it is also involved in the "looseness" of the t distribution.

Consider the estimation of variance. Again we can apply the principle of the confidence interval. The illustration of salinity calculation is quite appropriate here also. Many marine organisms are quite intolerant of changes in salinity, and the magnitude of variance could be especially inportant.

Suppose that many series of observations of salinity were made, and each time a sample of size 20 was used to estimate mean and variance. Both of these estimates are, of course, random variables, and each has its own distribution dependent on the true unknown population parameters and sampling accidents. On the basis of each series of observations we could obtain a new random variable

$$W = \frac{(n - 1)S^2}{\sigma^2} \tag{6.17}$$

where n is the sample size, S^2 the sample variance, and σ^2 the true unknown population variance. If we knew σ^2 we could work out the details of the distribution. This actually turns out to be the χ^2 distribution with $n - 1$ degrees of freedom. Values are presented in table A–5 in the appendix, where the number of degrees of freedom are ν.

Using the previous salinity data with samples of size 20, so that $\nu_2 = n - 1 = 19$ and assuming that our variance estimate is $S^2 = 1.400 - 1.960$, we now obtain a 95% CI for σ^2:

$$\chi^2_{.975} < W < \chi^2_{.025}$$

Note here that 2.5% of W values would be expected to be below $\chi^2_{.975}$ and 2.5% above $\chi^2_{.025}$, or

$$\chi^2_{.975} < \frac{(n - 1)S^2}{\sigma^2} < \chi^2_{.025}$$

Solving for σ^2 (It would be good practice for the reader to carry out the algebra here.)

$$\frac{(n - 1)S^2}{\chi^2_{.025}} < \sigma^2 < \frac{(n - 1)S^2}{\chi^2_{.975}}$$

For our problem

$$\frac{19 \cdot 1.960}{32.85} < \sigma^2 < \frac{19 \cdot 1.960}{8.91}$$

$$1.134 < \sigma^2 < 4.180$$

The 95%(CI) for σ^2 is

$$1.134 \text{ —— } 4.180.$$

Note that the interval is not symmetrical about the point of estimate. This is a consequence of the fact that the χ^2 distribution is bounded on the left by zero.

Analysis of Variance

Suppose we had t sets of observations taken from t different parts of a river. Were there no appreciable differences in the sites we would, a priori, expect no differences in our measure from one site to another. A very powerful technique for testing whether such a set of samples could reasonably be expected from a single, common universe is the method of analysis of variance. Basically this is a procedure for partitioning the total variance according to whether it comes from within samples or from differences between samples. Say we had t samples, each of size n:

	Samples							
	1	2	3		i		t
individual observations	X_{11}	X_{21}	X_{31}	...	X_{i1}	...	X_{t1}
	X_{12}	X_{22}	X_{32}	...	X_{i2}	...	X_{t2}
	X_{13}	X_{23}	X_{33}	...	X_{i3}	...	X_{t3}
	.	.	.		.		.
	.	.	.		.		.
	.	.	.		.		.
	X_{1j}	X_{2j}	X_{3j}	...	X_{ij}	...	X_{tj}
	.	.	.		.		.
	.	.	.		.		.
	.	.	.		.		.
	X_{1n}	X_{2n}	X_{3n}		X_{in}	...	X_{tn}

Here, then,

$$\bar{X}_1 = \frac{1}{n}\sum_{j=1}^{n} X_{1j},\ \bar{X}_2 = \frac{1}{n}\sum_{j=1}^{n} X_{2j},\ \ldots\ X_{i.} = \frac{1}{n}\sum_{j=1}^{n} X_{ij},\ \ldots\ X_{t.} = \frac{1}{n}\sum_{j=1}^{n} X_{tj}$$

And

$$\bar{\bar{X}} = \sum_{i=1}^{t}\sum_{j=1}^{n} X_{ij}$$

We can obtain measures that reflect the variability of the data in several different ways. The deviation of the individual observations from the various group means provides a measure of that component of the variance that is intrinsic to the various groups. This is generally termed the *error variance* or *within groups variance* and is in a sense the variability associated with sampling and observational error. This variation should be normally distributed with mean 0. Another source of variation can be assessed by looking at the difference between group means and the overall or grand mean. This would be related to the component of variability usually termed *among groups* or *between groups*. These two sources of variability must sum to the total variability—that which would be seen by looking at the deviation of all of the individual observations from the grand mean.

As with our earlier assessment of variability using variance as a measure, we here use a set of variance estimates that are related as in table

6–2. The column headed Source is self-evident. The categories represent a partitioning of the total variability into two components—that due to differences between the various groups, and that which is present within the groups. The column headed Degrees of Freedom specifies the number of values free to vary. Since there are t groups, and the total is fixed, only $t - 1$ can vary because the value of the last is then fixed as the remainder. Assuming that each group has n observations, there are $n - 1$ free to vary in each group, and there are t such groups (therefore, there are a total of nt observations when $nt - 1$ values are specified; the last is fixed as the remainder because the total is fixed). The Sum of Squares column presents the various manipulations necessary to compute the sum of squared deviations from the various means. In practice one usually computes only the among groups sum of squares (SS_A) and the Total and obtains the within groups sum of squares (SS_W) by subtraction. The Mean Square column is the analog of the variances and represents estimates of the values in the last column.

If there is no treatment effect, then $\Sigma_{i=1}^{n}\alpha_i^2$ should be 0 and $SS_A/(t - 1)$ should equal $SS_W/t(n - 1)$. We set $SS_A/(t - 1) = MS_A$ and $SS_W/t(n - 1) = MS_W$. Thus in the absence of treatment effect MS_A/MS_W would be expected to equal unity except for sampling variability. This provides the test for the null hypothesis—that is, "there are no treatment effects." Estimates of the same variance should be expected to differ only due to sampling, and the distribution of ratios of two such estimates will be influenced only by the size of the samples from which the estimates were obtained. In general, if S_1^2 is from a sample of size n_1, and S_2^2 is from a sample of size n_2, then

$$\frac{S_1^2}{S_2^2}$$

will be distributed as an F with $(n_1 - 1)$, $(n_2 - 1)$ degrees of freedom. The usual designation of degrees of freedom is ν_1 and ν_2, because only rarely is the number simply a count minus 1. For convenience, the F table (see table A–4 in the appendix) is only constructed for one side of the distribution, and the larger of the two estimates is always the numerator. Table A–4 is much abridged and only includes cut-off points for .10, .05, and .01, and only for a few values of ν_1 and ν_2. More values are presented for the denominator because the net effect of small differences is greatest here. Interpolation is practical with due caution.

Of considerable importance is a distinction to be made in the interpretation of treatment effects. We could have a situation in which the investigator, manipulating the experimental conditions, can fix the impact of treatment differences. Similarly, differences could be fixed between groups or treatment categories even though these differences are not di-

Table 6–2
Set of Variance Estimates

Source	*Degrees of Freedom*	*Sum of Squares*	*Mean Square*	*Estimate of*
Among groups	$t - 1$	$(1/n) \sum\left(\sum x_{ij}\right)^2 - (1/nt)\left(\sum\sum x_{ij}\right)^2$	$SS_A/t - 1$	$\sigma^2 + [n/(t - 1)] \sum \alpha_i^2$
Within groups	$t(n - 1)$	$\sum\sum x_{ij}^2 - (1/n) \sum\left(\sum x_{ij}\right)^2$	$SS_W/t\ (n - 1)$	σ^2
Total	$tn - 1$	$\sum\sum x_{ij}^2 - (1/nt) \left(\sum\sum x_{ij}\right)^2$		

rectly manipulated by the investigator. In either case the differences between treatments are due to fixed effects and are not random variables. Such a situation is conventionally referred to as a *Model I* or *fixed effect model*. We can express this by

$$X_{ij} = \mu + \alpha_i + \epsilon_{ij}$$

Here the individual observation X_{ij} is determined by a parametric value μ, the fixed treatment effect α_i, and finally a random component ϵ_{ij} that is distributed as a normal with mean 0 and an unknown variance σ^2. Here, in an experiment with untreated controls we could actually estimate values for α_i.

A different situation is encountered where the treatment effects are not fixed but rather are random variables, so that from one experiment to another the magnitude of treatment effects might vary considerably. Here the particular values are not as important as is the magnitude of the differences between the treatment effects. Here we could write

$$X_{ij} = \mu + A_i + \epsilon_{ij}$$

The individual observations are determined by the parametric μ, a treatment effect A_i that is a particular value from an unknown probability distribution, and finally the random component ϵ_{ij}. Just what treatment effects and hence values A_i included in this experiment that are not manipulated by the experimenter is not as important as is the question of simply whether the various A_i values are alike or differ from one another. This situation is conventionally referred to as a *Model II* or *random effect model*. Because of this distinction, there is a difference in the variance-like components involved in the variance ratio (F tests) for the significance of treatment effects.

In Model I,

$$F \approx \frac{S^2 + \dfrac{n}{t-1}\alpha_i^2}{S^2}$$

In Model II,

$$F \approx \frac{S^2 + n\sigma_A^2}{S^2}$$

Where σ_A^2 is the variance of the treatment-affected distribution. Now we work through an example of analysis of variance.

Example 6.5. Suppose we have three sets of dissolved oxygen observations taken from points in a river just downstream from confluences with three tributaries. Our interest is in detecting differences on the effect of tributary flow—not in general—but rather specifically from these three tributaries. These do not represent a random sample of all possible tributaries, but are chosen much as one might choose dosage levels or medications in a drug experiment. Clearly, this is a Model I situation:

	Tributaries 1	2	3	
	5.4	5.1	5.5	
Dissolved oxygen observations	5.2	4.6	5.0	
	4.7	5.2	5.4	
	5.6	4.8	4.8	
	5.1	4.6	5.7	
	5.5	5.4	5.6	
	5.1	5.0	5.3	
	5.3	4.9	5.5	
x_{ij}	41.9	39.6	42.8	$x_{ij} = 124.3$
x_{ij}^2	220.01	196.58	229.64	$x_{ij}^2 = 646.23$
	5.2357	4.9500	5.3500	$\bar{\bar{x}}.. = 5.1792$

In this illustration our explicit hypothesis is that the means are equal. The implicit hypothesis (the one we actually test) is that the two independent estimates of the variance are the same. We arbitrarily choose α to be .05. An error of the first kind here (as described in the section on hypothesis testing) would be to say that the means differ when actually they do not.

We could expect to make such an error 5% of the time when the means are alike:

Preliminary Calculations

Source	*Sum of Squares*	*Degrees of Freedom*	*Mean Square*	*F*
among groups	644.4513 − 643.7704 = .6809	2	.3405	4.0195
within groups	646.2300 − 644.4513 = 1.7787	21	.0847	
total	626.2300 − 643.7704 = 2.4596	23		

From table A–4 in the appendix we find that $F_{.05}(2,21) = 3.4668$. Therefore, since our calculated F exceeds the table value, we reject the hypothesis tested.

The reader should follow through on all of these calculations and make sure that the various components of the analysis of variance table are clear. It should be pointed out that this test of an hypothesis about means can be executed without even calculating the means at all. In this case it is interesting to note that if we were to calculate the means and obtain 95% confidence intervals for all of the means, they would all overlap.

In our table we have suggested that if there is any difference between groups, it will show up in the variability in the form

$$\frac{n}{t - 1} \alpha_i^2$$

We are assuming here that there may be some unknown but fixed group of treatment effects. In more sophisticated and thorough treatments of analysis of variance, this type of model will be referred to as a "fixed effects" or Model I situation. It is also possible that differences between groups could be random effects rather than fixed, and that the specific values which enter the experiment are a sample selection from the possible array of effects. This would be a Model II, and the treatment would appear as $n\sigma_m^2$, where σ_m^2 would reflect the variance of the random treatment effect. There is a meaningful difference here, and the reader is cautioned to check further before making application of the analysis of variance technique in any serious situation.

A few assumptions must be made about the random variables involved to apply analysis of variance techniques. Although the test is fairly robust in the sense that minor violations of assumptions are not likely to lead to grossly erroneous results, one must still be aware of the assumptions. Sometimes it is necessary to transform the data to bring it into conformity, but often the violations of assumptions are such that the technique is simply not applicable.

Assumptions

(A) Variances within each group or subgroup must be the same. The uniformity of variances can be tested, and should be if there is doubt. Suitable tests can be found in any standard text.
(B) The random variables within groups must be normally distributed. Visual check for grossly skewed distributions will often alert the investigator. Often a distribution such as a Poisson with a very low mean or a distribution with a logarithmic form may be amenable to a logarithmic transformation.

Analysis of variance is a very powerful technique. Our simple illustration involved only a single variable of classification, but the technique is applicable with several variables of classification. If each treatment combination is replicated, it is possible to test for interactions of factors as well as for the significance of the individual factors. Analysis of variance is closely tied to a whole discipline of experimental design. Sophisticated design technique permits extremely efficient use of experimental resources with the retention of powerful testing procedures.

Appendix Statistical Tables (A1 to A4)

Table A–1
The Standard Normal Distribution

Z	$f(Z)$	$P(Z)$
0	.3989	.5000
.1	.3970	.4602
.2	.3910	.4207
.3	.3814	.3821
.4	.3683	.3446
.5	.3521	.3085
.6	.3332	.2743
.7	.3123	.2420
.8	.2897	.2119
.9	.2661	.1841
1.0	.2420	.1587
1.1	.2179	.1357
1.2	.1942	.1151
1.3	.1714	.0968
1.4	.1497	.0808
1.5	.1295	.0668
1.6	.1109	.0548
1.7	.0940	.0446
1.8	.0790	.0359
1.9	.0656	.0287
2.0	.0540	.0228
2.5	.0175	.0062
3.0	.0044	.0013
4.0	.0001	.0000
1.282	.1754	.10
1.645	.1031	.05
1.960	.0584	.025
2.326	.0267	.01
2.576	.0145	.005
2.807	.0078	.0025
3.090	.0034	.001

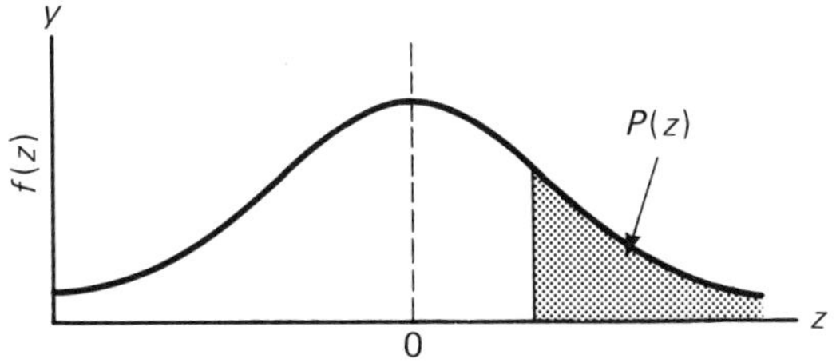

$$f(z) = \frac{1}{\sqrt{2\pi}} \exp(-z^2/2) \qquad P(z) = \frac{1}{\sqrt{2\pi}} \int_z^\infty \exp(-u^2/2)\, du$$

Source: Table derived by computer approximation (E. R. Rich).

Table A–2
Critical Values of Student's *t* Distribution

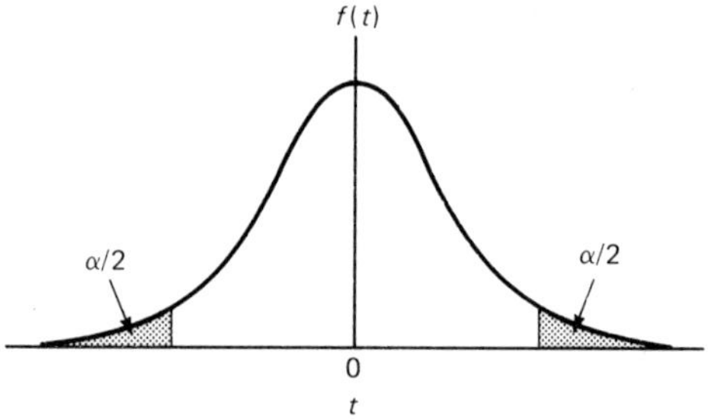

$\nu \backslash \alpha$	*0.1*	*0.05*	*0.02*	*0.01*	*0.001*	$\alpha\nu$
1	6.314	12.706	31.821	63.657	636.619	1
2	2.290	4.303	6.965	9.925	31.598	2
3	2.353	3.182	5.451	5.841	12.924	3
4	2.132	2.776	3.747	4.604	8.610	4
5	2.015	2.571	3.365	4.032	6.869	5
6	1.943	2.447	3.143	3.707	5.959	6
7	1.895	2.365	2.998	3.499	5.408	7
8	1.660	2.306	2.896	3.355	5.041	8
9	1.833	2.262	2.821	3.250	4.781	9
10	1.812	2.228	2.764	3.169	4.587	10
11	1.796	2.201	2.718	3.106	4.437	11
12	1.782	2.179	2.681	3.055	4.318	12
13	1.771	2.160	2.650	3.012	4.221	13
14	1.761	2.145	2.624	2.977	4.140	14
15	1.753	2.131	2.602	2.947	4.073	15
16	1.746	2.120	2.583	2.921	4.015	16
17	1.740	2.110	2.567	2.898	3.965	17
18	1.734	2.101	2.552	2.878	3.922	18
19	1.729	1.093	2.539	2.861	3.883	19
20	1.725	2.086	2.528	2.845	3.850	20
21	1.721	2.080	2.518	2.831	3.819	21
22	1.717	2.074	2.508	2.819	3.792	22
23	1.714	2.069	2.500	2.807	3.767	23
24	1.711	2.064	2.492	2.797	3.745	24
25	1.708	2.060	2.485	2.787	3.725	25
26	1.706	2.056	2.479	2.779	3.707	26
27	1.703	2.052	2.473	2.771	3.690	27
28	1.701	2.048	2.467	2.763	3.674	28
29	1.699	2.045	2.462	2.756	3.659	29
30	1.697	2.042	2.457	2.750	3.646	30
40	1.684	2.021	2.423	2.704	3.551	40
60	1.671	2.000	2.390	2.660	3.469	60
120	1.658	1.980	2.358	2.617	3.373	120
∞	1.645	1.960	2.326	2.576	3.291	∞

Table A–3
Critical Values of the Chi-Square Distribution

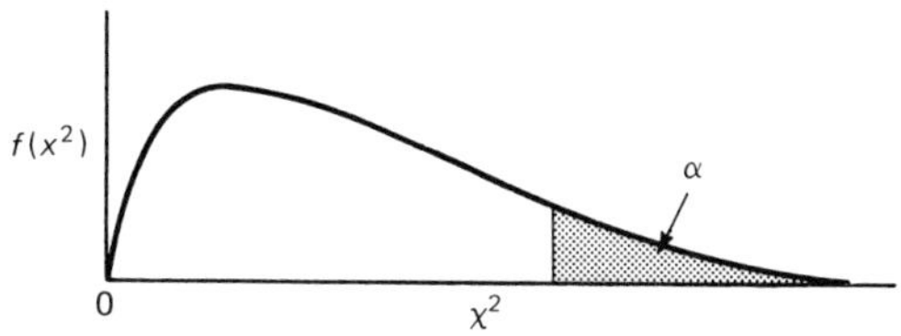

ν \ α	*.01*	*0.05*	*0.025*	*0.01*	*0.005*	ν
1	2.706	3.841	5.024	6.635	7.879	1
2	4.605	5.991	7.378	9.210	10.597	2
3	6.251	7.815	9.348	11.345	12.838	3
4	7.779	9.488	11.143	13.277	14.860	4
5	9.236	11.070	12.832	15.086	16.750	5
6	10.645	12.592	14.449	16.812	18.548	6
7	12.017	14.067	16.013	18.475	20.278	7
8	13.362	15.507	17.535	20.090	21.955	8
9	14.684	16.919	19.023	21.666	23.589	9
10	15.987	18.307	20.483	23.209	25.188	10
11	17.275	19.675	21.920	24.725	26.757	11
12	18.549	21.026	23.337	26.217	28.300	12
13	19.812	22.362	24.736	27.688	29.819	13
14	21.064	23.685	26.119	29.141	31.319	14
15	22.307	24.996	27.488	30.578	32.801	15
16	23.542	26.296	28.845	32.000	34.267	16
17	24.769	27.587	30.191	33.409	35.718	17
18	25.989	28.869	31.526	34.805	37.156	18
19	27.204	30.144	32.852	36.191	38.582	19
20	28.412	31.410	34.170	37.566	39.997	20
21	29.615	32.670	35.479	38.932	41.401	21
22	30.813	33.924	36.781	40.289	42.796	22
23	32.007	35.172	38.076	41.638	44.181	23
24	33.196	36.415	39.364	42.980	45.558	24
25	34.382	37.652	40.646	44.314	46.928	25
26	35.563	38.885	41.923	45.642	48.290	26
27	36.741	40.113	43.194	46.963	49.645	27
28	37.916	41.337	44.461	48.278	50.993	28
29	39.088	42.557	45.722	49.588	52.336	29
30	40.256	43.773	46.679	50.892	63.672	30
31	41.422	44.985	48.232	52.192	55.003	31
32	42.585	48.194	49.481	53.486	56.329	32
33	43.745	47.400	50.725	54.776	57.649	33
34	44.903	48.602	51.966	56.061	58.964	34
35	46.059	49.802	53.203	57.342	60.275	35

Table A–3 (continued)

$\nu \backslash \alpha$	*.01*	*0.05*	*0.025*	*0.01*	*0.005*	ν
36	47.212	50.998	54.437	58.619	61.582	36
37	48.363	52.192	55.668	59.893	62.884	37
38	49.513	53.384	56.896	61.162	64.182	38
39	50.660	54.572	58.120	62.428	65.476	39
40	51.805	55.758	59.342	63.691	66.766	40
41	52.949	56.942	60.561	64.950	68.053	41
42	54.090	58.124	61.777	66.206	69.336	42
43	55.230	59.304	62.990	67.460	70.616	43
44	56.369	60.481	64.202	68.710	71.893	44
45	57.505	61.656	65.410	69.957	73.166	45
46	58.641	62.830	66.617	71.202	74.437	46
47	59.774	64.001	67.821	72.443	75.704	47
48	60.907	65.171	69.023	73.683	76.969	48
49	62.038	66.339	70.222	74.920	78.231	49
50	63.167	67.505	71.420	76.154	79.490	50

Table A–4
Critical Values of the *F* Distribution

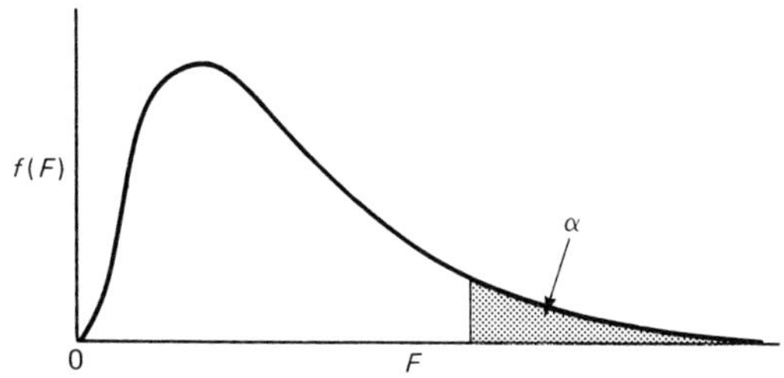

ν_1 *(degrees of freedom of numerator mean square)*

ν_2 (degrees of freedom of denominator mean square)	α	*1*	*2*	*3*	*4*	*5*	*6*	*7*	*8*	*9*	*10*	*11*
	.10	39.9	49.5	53.6	55.8	57.2	58.2	58.9	59.4	59.9	60.2	60.5
1	.05	161	199	216	225	230	234	237	239	241	241	243
	.01	4050	5000	5400	5620	5760	5860	5930	5980	6020	6060	6080
	.10	8.53	9.00	9.16	9.24	9.29	9.33	9.35	9.37	9.38	9.39	9.40
2	.05	18.5	19.0	19.2	19.2	19.3	19.3	19.4	19.4	19.4	19.4	19.4
	.01	98.5	99.0	99.2	99.2	99.3	99.3	99.4	99.4	99.4	99.4	99.4
	.10	5.54	5.46	5.39	5.34	5.31	5.28	5.27	5.25	5.24	5.23	5.22
3	.05	10.1	9.55	9.28	9.12	9.01	8.94	8.89	8.85	8.81	8.79	8.76
	.01	34.1	30.8	29.5	28.7	28.2	27.9	27.7	27.5	27.3	27.2	27.1
	.10	4.54	4.32	4.19	4.11	4.05	4.01	3.98	3.95	3.94	3.92	3.91
4	.05	7.71	6.94	6.59	6.39	8.26	6.16	6.09	6.04	6.00	5.96	5.93
	.01	21.2	18.0	16.7	16.0	15.5	15.2	15.0	14.8	14.7	14.5	14.4
	.10	4.06	3.78	3.62	3.52	3.45	3.40	3.37	3.34	3.32	3.30	3.28
5	.05	6.61	5.79	5.41	5.19	5.05	4.95	4.88	4.82	4.77	4.74	4.71
	.01	16.3	13.3	12.1	11.4	11.0	10.7	10.5	10.3	10.2	10.1	9.99
	.10	3.78	3.46	3.29	3.18	3.11	3.05	3.01	2.98	2.96	2.94	2.92
6	.05	5.99	5.14	4.76	4.53	4.39	4.28	4.21	4.15	4.10	4.06	4.03
	.01	13.7	10.9	9.78	9.15	8.75	8.47	8.26	8.10	7.98	7.87	7.79
	.10	3.59	3.26	3.07	2.98	2.88	2.83	2.78	2.75	2.72	2.70	2.68
7	.05	5.59	4.74	4.35	4.12	3.97	3.87	3.77	3.73	3.68	3.64	3.60
	.01	12.2	9.55	8.45	7.85	7.46	7.19	6.99	6.84	6.72	6.62	6.54
	.10	3.46	3.11	2.92	2.81	2.73	2.67	2.62	2.59	2.56	2.54	2.52
8	.05	5.32	4.46	4.07	3.84	3.69	3.58	3.50	3.44	3.39	3.35	3.31
	.01	11.3	8.65	7.59	7.01	6.63	6.37	6.18	6.03	5.91	5.81	5.73
	.10	3.36	3.01	2.81	2.69	2.61	2.55	2.51	2.47	2.44	2.42	2.40
9	.05	5.12	4.26	3.86	3.63	3.48	3.37	3.29	3.23	3.18	3.14	3.10
	.01	10.6	8.02	6.99	6.42	6.06	5.80	5.61	5.47	5.35	5.26	5.18
	.10	3.29	2.92	2.73	2.61	2.52	2.46	2.41	2.38	2.35	2.32	2.30
10	.05	4.96	4.10	3.71	3.48	3.33	3.22	3.14	3.07	3.02	2.98	2.94
	.01	10.0	7.56	6.55	5.99	5.64	5.39	5.20	5.06	4.94	4.85	4.77

Table A–4
(continued)

ν_2 (degrees of freedom of denominator mean square)	ν_1 (degrees of freedom of numerator mean square)											
	α	*1*	*2*	*3*	*4*	*5*	*6*	*7*	*8*	*9*	*10*	*11*
	.10	3.23	2.86	2.66	2.54	2.45	2.39	2.34	2.30	2.27	2.25	2.23
11	.05	4.84	3.98	3.59	3.36	3.20	3.09	3.01	2.95	2.90	2.85	2.82
	.01	9.65	7.21	6.22	5.67	5.32	5.07	4.89	4.74	4.63	4.54	4.46
	.10	3.18	2.81	2.61	2.48	2.39	2.33	2.28	2.24	2.21	2.19	2.17
12	.05	4.75	3.89	3.49	3.26	3.11	3.00	2.91	2.85	2.80	2.75	2.72
	.01	9.33	6.93	5.95	5.41	5.06	4.82	4.64	4.50	4.39	4.30	4.22
	.10	3.14	2.76	2.56	2.43	2.35	2.28	2.23	2.20	2.16	2.14	2.12
13	.05	4.67	3.81	3.41	3.18	3.03	2.92	2.83	2.77	2.71	2.67	2.64
	.01	9.07	6.70	5.74	4.21	4.86	4.62	4.44	4.30	4.19	4.10	4.03
	.10	3.10	2.73	2.52	2.39	2.31	2.24	2.19	2.45	1.42	2.10	2.07
14	.05	4.60	3.74	3.34	3.11	2.96	2.85	2.76	2.70	2.65	2.60	2.57
	.01	8.86	6.51	5.56	5.04	4.69	4.46	4.28	4.14	4.03	3.94	3.86
	.10	3.07	2.70	2.49	2.36	2.27	2.21	2.16	2.12	2.09	2.54	2.51
15	.05	4.54	3.68	3.29	3.06	2.90	2.79	2.71	2.64	2.59	2.06	2.04
	.01	8.68	6.36	5.42	4.89	4.56	4.32	4.14	4.00	3.89	3.80	3.73
	.10	3.05	2.67	2.46	2.33	2.24	2.18	2.13	2.09	2.06	2.03	2.01
16	.05	4.49	3.63	3.24	3.01	2.85	2.74	2.66	2.59	2.54	2.49	2.46
	.01	8.53	6.23	5.29	4.77	4.44	4.20	4.03	3.89	3.78	3.69	3.62
	.10	3.03	2.64	2.44	2.31	2.22	2.15	2.10	2.06	2.03	2.00	1.98
17	.05	4.45	3.59	3.20	2.96	2.91	2.70	3.61	2.55	2.49	2.45	2.41
	.01	8.40	6.11	5.19	4.67	4.34	4.10	3.93	3.79	3.68	3.59	3.52
	.10	3.01	2.62	2.42	2.29	2.20	2.13	2.08	2.04	2.00	1.98	1.95
18	.05	4.41	3.55	3.16	2.93	2.77	2.66	2.58	2.51	2.46	2.41	2.37
	.01	8.28	6.01	5.09	4.58	4.25	4.01	3.84	3.71	3.60	3.51	3.43
	.10	2.99	2.61	2.40	7.27	2.18	2.11	2.06	2.02	1.98	1.96	1.93
19	.05	4.38	3.52	3.13	2.90	2.74	2.63	2.54	2.48	2.42	2.38	2.34
	.01	8.19	5.93	5.01	4.50	4.17	3.94	3.77	3.63	3.52	3.43	3.56
	.10	2.97	2.59	2.38	2.25	2.16	2.09	2.04	2.00	1.96	1.94	1.91
20	.05	4.35	3.49	3.10	2.87	2.71	2.60	2.51	2.45	2.39	2.35	2.31
	.01	8.10	5.85	4.94	4.43	4.10	3.87	3.70	3.56	3.46	3.37	3.29
	.10	2.96	2.57	2.36	2.23	2.14	2.08	2.02	1.98	1.95	1.92	1.90
21	.05	4.32	3.47	3.07	2.84	2.68	2.57	2.49	2.42	2.37	2.32	2.28
	.01	8.02	5.75	4.87	4.37	4.04	3.81	3.64	3.51	3.40	3.31	3.24
	.10	2.95	2.56	2.35	2.22	2.13	2.06	2.01	1.97	1.93	1.90	1.88
22	.05	4.30	3.44	3.05	2.82	2.66	2.55	2.46	2.40	2.39	2.30	2.26
	.01	7.95	5.72	4.82	4.31	3.99	3.76	3.59	3.45	3.35	3.26	3.18

Table A–4
(continued)

ν_2 (degrees of freedom of denominator mean square)	α	ν_1 (degrees of freedom of numerator mean square)										
		1	*2*	*3*	*4*	*5*	*6*	*7*	*8*	*9*	*10*	*11*
	.10	2.94	2.55	2.31	2.25	2.11	1.05	1.99	1.95	1.92	1.89	1.87
23	.05	4.28	3.42	3.03	2.80	2.84	2.53	2.44	2.37	2.32	2.27	2.24
	.01	7.88	5.66	4.76	4.26	3.94	3.71	3.54	3.41	3.30	3.21	3.14
	.10	2.93	2.54	2.33	2.19	2.10	2.04	1.98	1.94	1.91	1.88	1.85
24	.05	4.26	3.40	3.01	2.78	2.62	2.51	2.42	2.36	2.30	2.25	2.22
	.01	7.82	5.61	4.72	4.22	3.90	3.67	3.50	3.36	3.26	3.17	3.09
	.10	2.92	2.53	2.32	2.18	2.09	2.02	1.97	1.93	1.89	1.87	1.84
25	.05	4.24	3.39	2.99	2.76	2.60	2.49	2.40	2.34	2.28	2.24	2.20
	.01	7.77	5.57	4.68	4.18	3.86	3.63	3.46	3.32	3.22	3.13	3.06
	.10	2.91	2.52	2.31	2.17	2.08	2.01	1.96	1.92	1.88	1.86	1.83
26	.05	4.23	3.37	2.98	2.74	2.59	2.47	2.39	2.32	2.27	2.22	2.18
	.01	7.72	5.53	4.64	4.14	3.82	3.59	3.42	3.29	3.18	3.09	3.02
	.10	2.90	2.51	2.30	2.17	2.07	2.00	1.95	1.91	1.87	1.85	1.82
27	.05	4.21	3.35	2.96	2.73	2.57	2.46	2.37	2.31	2.25	2.20	2.17
	.01	7.68	5.49	4.60	4.11	3.78	3.56	3.39	3.26	3.45	3.06	2.99
	.01	2.89	2.50	2.29	2.16	2.06	2.00	1.94	1.90	1.87	1.84	1.81
28	.05	4.20	3.34	2.95	2.71	2.56	2.45	2.36	2.29	2.24	2.19	2.15
	.01	7.64	5.45	4.57	4.07	3.75	3.53	3.36	3.23	3.12	3.03	2.96
	.10	2.89	2.50	2.28	2.15	2.06	1.99	1.93	1.89	2.86	1.83	1.80
29	.05	4.18	3.33	2.93	2.70	2.55	2.43	2.35	2.28	2.22	2.18	2.14
	.01	7.60	5.42	4.54	4.04	3.73	3.50	3.33	3.20	3.09	3.00	2.93
	.10	2.88	2.49	2.28	2.14	2.05	1.98	1.93	1.88	1.85	1.82	1.79
30	.05	4.17	3.32	2.92	2.69	2.53	2.42	2.33	2.27	2.21	2.16	2.13
	.01	7.56	5.39	4.51	4.02	3.70	3.47	3.30	3.17	3.07	2.98	2.90
	.10	2.84	2.44	2.23	2.09	2.00	1.93	1.87	1.83	1.79	1.76	1.74
40	.05	4.08	3.23	2.84	2.61	2.45	2.34	2.25	2.18	2.12	2.08	2.04
	.01	7.31	5.18	4.31	3.83	3.51	3.29	3.12	2.99	2.89	2.80	2.73
	.10	2.79	2.39	2.18	2.04	1.95	1.87	1.82	1.77	1.74	1.71	1.95
60	.05	4.00	3.15	2.76	2.53	2.37	2.25	2.17	2.10	2.04	1.99	1.95
	.01	7.08	4.98	4.13	3.65	3.34	3.12	2.95	2.82	2.72	2.63	2.56
	.10	2.75	2.35	2.13	1.99	1.90	1.82	1.77	1.72	1.68	1.65	1.63
120	.05	3.92	3.07	2.68	2.45	2.29	2.17	2.09	2.02	1.96	1.91	1.87
	.01	6.85	4.79	3.95	3.48	3.17	2.96	2.79	2.66	2.56	2.47	2.40
	.10	2.71	2.30	2.08	1.94	1.85	1.77	1.72	1.67	1.63	1.60	1.57
∞	.05	3.84	3.00	2.60	2.37	2.21	2.10	2.01	1.94	1.88	1.83	1.79
	.01	6.63	4.61	3.78	3.32	3.02	2.80	2.64	2.51	2.41	2.32	2.25

Table A–4
(continued)

ν_2 (degrees of freedom of denominator mean square)	α	*ν_1 (degrees of freedom of numerator mean square)* 12	15	20	24	30	40	50	60	120	∞	
	.10	60.7	61.2	61.7	62.0	62.3	62.5	62.7	62.8	63.1	63.3	.10
1	.05	244	246	248	249	250	251	252	252	253	254	.05
	.01	6110	6160	6210	6230	6260	6290	6300	6310	6340	6370	.01
	.10	9.41	9.42	9.44	9.45	9.46	9.47	9.47	9.47	9.48	9.49	.10
2	.05	19.4	19.4	19.4	19.5	19.5	19.5	19.5	19.5	19.5	19.5	.05
	.01	99.4	99.4	99.4	99.5	99.5	99.5	99.5	99.5	99.5	99.5	.01
	.10	5.22	5.20	5.18	5.18	5.17	5.16	5.15	5.15	5.14	5.13	.10
3	.05	8.74	8.70	8.66	8.64	8.62	8.59	8.58	8.57	8.55	8.53	.05
	.01	27.1	26.9	26.7	26.6	26.5	26.4	26.3	26.3	26.2	26.1	.01
	.10	3.90	3.87	3.84	3.83	3.82	3.80	3.79	3.79	3.78	3.76	.10
4	.05	5.91	5.86	5.80	5.77	5.75	5.72	5.70	5.69	5.66	5.63	.05
	.01	14.4	14.2	14.0	13.9	13.8	13.7	13.7	13.7	13.6	13.5	.01
	.10	3.27	3.24	3.21	3.19	3.17	3.16	3.15	3.14	3.12	3.10	.10
5	.05	4.68	4.62	4.56	4.53	4.50	4.46	4.44	4.43	4.40	4.36	.05
	.01	9.89	9.72	9.55	9.47	9.38	9.29	9.24	9.20	9.11	9.02	.01
	.10	2.90	2.87	2.84	2.82	2.80	2.78	2.77	2.76	2.74	2.72	
6	.05	4.00	3.94	3.87	3.84	3.81	3.77	3.75	3.74	3.70	3.67	
	.01	7.72	7.56	7.40	7.31	7.23	7.14	7.09	7.06	6.97	6.88	
	.10	2.67	2.63	2.59	2.58	2.56	2.54	2.52	2.51	2.49	2.47	
7	.05	3.57	3.51	3.44	3.41	3.38	3.34	3.32	3.30	3.27	3.23	
	.01	6.47	6.31	6.16	6.07	5.99	5.91	5.86	5.82	5.74	5.65	
	.10	2.50	2.46	2.42	2.40	2.38	2.36	2.35	2.34	2.32	2.29	
8	.05	3.28	3.22	3.15	3.12	3.08	3.04	3.02	3.01	2.97	2.93	
	.01	5.67	5.52	5.36	5.28	5.20	5.12	5.07	5.03	4.95	4.86	
	.10	2.38	2.34	2.30	2.28	2.25	2.23	2.22	2.21	2.18	2.16	
9	.05	3.07	3.01	2.94	2.90	2.86	2.83	2.81	2.79	2.75	2.71	
	.01	5.11	4.96	4.81	4.73	4.65	4.57	4.52	4.48	4.40	4.31	
	.10	2.28	2.24	2.20	2.18	2.16	2.13	2.12	2.11	2.08	2.06	
10	.05	2.91	2.85	2.77	2.74	2.70	2.66	2.64	2.62	2.58	2.54	
	.01	4.71	4.56	4.41	4.33	4.25	4.17	4.12	4.08	4.00	3.91	
	.10	2.21	2.17	2.12	2.10	2.08	2.05	2.04	2.03	2.00	1.97	.10
11	.05	2.79	2.72	2.65	2.61	2.57	2.53	2.51	2.49	2.45	2.40	.05
	.01	4.40	4.25	4.10	4.02	3.94	3.86	3.81	3.78	3.69	3.60	.01
	.10	2.15	2.11	2.06	2.04	2.01	1.99	1.97	1.96	1.93	1.90	.10
12	.05	2.69	2.62	2.54	2.51	2.47	2.43	2.40	2.38	2.34	2.30	.05
	.01	4.16	4.01	3.86	3.78	3.70	3.62	3.57	3.54	3.45	3.36	.01
	.10	2.10	2.05	2.01	1.98	1.96	1.93	1.92	1.90	1.88	1.85	.10
13	.05	2.60	2.53	2.46	2.42	2.38	2.34	2.31	2.30	2.25	2.21	.05
	.01	3.96	3.82	3.66	3.59	3.51	3.43	3.37	3.34	3.25	3.17	.01

Table A–4 (continued)

ν_2 (degrees of freedom of denominator mean square)	α	12	15	20	24	30	40	50	60	120	∞	
		ν_1 (degrees of freedom of numerator mean square)										
	.10	2.05	2.01	1.96	1.94	1.91	1.89	1.87	1.86	1.83	1.80	.10
14	.05	2.53	2.46	2.39	2.35	2.31	2.27	2.24	2.22	2.18	2.13	.05
	.01	3.80	3.66	3.51	3.43	3.35	3.27	3.21	3.18	3.09	3.00	.01
	.10	2.48	1.97	1.92	1.90	1.87	1.85	1.83	1.82	1.79	1.76	.10
15	.05	2.02	2.40	2.33	2.39	2.25	2.20	2.18	2.16	2.11	2.07	.05
	.01	3.67	3.52	3.37	3.29	3.21	3.13	3.08	3.05	2.96	2.87	.01
	.10	1.99	1.94	1.89	1.87	1.84	1.81	1.79	1.78	1.75	1.72	.10
16	.05	2.42	2.35	2.28	2.24	2.19	2.15	2.12	2.11	2.06	2.01	.05
	.01	3.55	3.41	3.26	3.18	3.10	3.02	2.97	2.93	2.84	2.75	.01
	.10	1.96	1.91	1.86	1.84	1.81	1.78	1.76	1.75	1.72	1.69	.10
17	.05	2.38	2.31	2.23	2.19	2.15	2.10	2.08	2.06	2.01	1.96	.05
	.01	3.46	3.31	3.16	3.08	3.00	2.92	2.87	2.83	2.75	2.65	.01
	.10	1.93	1.89	1.84	1.81	1.78	1.75	1.74	1.72	1.69	1.66	.10
18	.05	2.34	2.27	2.19	2.15	2.11	2.06	2.04	2.02	1.97	1.92	.05
	.01	3.37	3.23	3.08	3.00	2.92	2.84	2.78	2.75	2.66	2.57	.01
	.10	1.91	1.86	1.81	1.79	1.76	1.73	1.71	1.70	1.67	1.63	.10
19	.05	2.31	2.23	2.16	2.11	2.07	2.03	2.00	1.98	1.93	1.88	.05
	.01	3.30	3.15	3.00	2.92	2.84	2.76	2.71	2.67	2.58	2.49	.01
	.10	1.89	1.84	1.79	1.77	1.74	1.71	1.69	1.68	1.64	1.61	.10
20	.05	2.28	2.20	2.12	2.08	2.04	1.99	1.97	1.95	1.90	1.84	.05
	.01	3.23	3.09	2.94	2.86	2.78	2.69	2.64	2.61	2.52	2.42	.01
	.10	1.87	1.83	1.78	1.75	1.72	1.69	1.67	1.66	1.62	1.59	
21	.05	2.25	2.18	2.10	2.05	2.01	1.96	1.94	1.92	1.87	1.81	
	.01	3.17	3.03	2.88	2.80	2.72	2.64	2.58	2.55	2.46	2.36	
	.10	1.86	1.81	1.76	1.73	1.70	1.67	1.65	1.64	1.60	1.57	
22	.05	2.23	2.15	2.07	2.03	1.98	1.94	1.91	1.89	1.84	1.78	
	.01	3.12	2.98	2.83	2.75	2.67	2.58	2.53	2.50	2.40	2.31	
	.10	1.85	1.80	1.74	1.72	1.69	1.66	1.64	1.62	1.59	1.55	
23	.05	2.20	2.13	2.05	2.01	1.96	1.91	1.89	1.86	1.81	1.76	
	.01	3.07	2.93	2.78	2.70	2.62	2.54	2.48	2.45	2.35	2.26	
	.10	1.83	1.78	1.73	1.70	1.67	1.64	1.62	1.61	1.57	1.53	
24	.05	2.18	2.11	2.03	1.98	1.94	1.89	1.86	1.84	1.79	1.73	
	.01	3.03	2.89	2.74	2.66	2.58	2.49	2.44	2.40	2.31	2.21	
	.10	1.82	1.77	1.72	1.69	1.66	1.63	1.61	1.59	1.56	1.52	
25	.05	2.16	2.09	2.01	1.96	1.92	1.87	1.84	1.82	1.77	1.71	
	.01	2.99	2.85	2.70	2.62	2.54	2.45	2.40	2.36	2.27	2.17	
	.10	1.81	1.76	1.71	1.68	1.65	1.61	1.59	1.58	1.54	1.50	
26	.05	2.15	2.07	1.99	1.95	1.90	1.85	1.82	1.80	1.75	1.69	
	.01	2.96	2.81	2.66	2.58	2.50	2.42	2.36	2.33	2.23	2.13	

Table A–4 (continued)

ν_2 (degrees of freedom of denominator mean square)	α	12	15	20	24	30	40	50	60	120	∞
		ν_1 (degrees of freedom of numerator mean square)									
27	.10	1.80	1.75	1.70	1.67	1.64	1.60	1.58	1.57	1.53	1.49
	.05	2.13	2.06	1.97	1.93	1.88	1.84	1.81	1.79	1.73	1.67
	.01	2.93	2.78	2.63	2.55	2.47	2.38	2.33	2.29	2.20	2.10
28	.10	1.79	1.74	1.69	1.66	1.63	1.59	1.57	1.56	1.52	1.48
	.05	2.12	2.04	1.96	1.91	1.87	1.82	1.79	1.77	1.71	1.65
	.01	2.90	2.75	2.60	2.52	2.44	2.35	2.30	2.26	2.17	2.06
29	.10	1.78	1.73	1.68	1.65	1.62	1.58	1.56	1.55	1.51	1.47
	.05	2.10	2.03	1.94	1.90	1.85	1.81	1.78	1.75	1.70	1.64
	.01	2.87	2.73	2.57	2.49	2.41	2.33	2.27	2.23	2.14	2.03
30	.10	1.77	1.72	1.67	1.64	1.61	1.57	1.55	1.54	1.50	1.46
	.05	2.09	2.01	1.93	1.89	1.84	1.79	1.76	1.74	1.68	1.62
	.01	2.84	2.70	2.55	2.47	2.39	2.30	2.25	2.21	2.11	2.01
40	.10	1.74	1.66	1.61	1.57	1.54	1.51	1.48	1.47	1.42	1.38
	.05	2.04	1.92	1.84	1.79	1.74	1.69	1.66	1.64	1.58	1.51
	.01	2.66	2.52	2.37	2.29	2.20	2.11	2.06	2.02	1.92	1.80
60	.10	1.66	1.60	1.54	1.51	1.48	1.44	1.41	1.40	1.35	1.29
	.05	1.92	1.84	1.75	1.70	1.65	1.59	1.56	1.53	1.47	1.39
	.01	2.50	2.35	2.20	2.12	2.03	1.94	1.88	1.84	1.73	1.60
120	.10	1.60	1.55	1.48	1.45	1.41	1.37	1.34	1.32	1.26	1.19
	.05	1.83	1.75	1.66	1.61	1.55	1.50	1.46	1.43	1.35	1.25
	.01	2.34	2.19	2.03	1.95	1.86	1.76	1.70	1.66	1.53	1.38
∞	.10	1.55	1.49	1.42	1.38	1.34	1.30	1.26	1.24	1.17	1.00
	.05	1.75	1.67	1.57	1.52	1.46	1.39	1.35	1.32	1.22	1.00
	.01	2.18	2.04	1.88	1.79	1.70	1.59	1.52	1.47	1.32	1.00

Index

Index

About the Authors

Thomas D. Waite received the B. S. and M. S. in civil engineering at Northeastern University, and the S. M. and Ph. D. in environmental engineering from Harvard. He is associate professor of environmental health engineering at Northwestern University. His interests deal with microbiological and chemical processes regulating water quality.

Neil J. Freeman received the B. S. and M. S. in civil engineering at the University of Miami, and the Ph. D. in theoretical and applied mechanics at Northwestern University. He is associate professor of civil engineering at the University of Miami. His interests lie in the area of modeling environmental processes.